Contents

Language

> Climate disruption is already having diverse and wide ranging impacts on Ireland's environment, society, economic and natural resources. The Climate Action Plan sets out an ambitious course of action over the coming years to address this issue.

This was the official text used to hail the release on the 17 June 2019 of the Irish Government's *Climate Action Plan 2019*. The plan is indeed ambitious covering Electricity, Buildings, Transport, Agriculture, Enterprise and Services, Waste and the Circular Economy. The authors of the plan know that evidence is never enough. Citizens need to be persuaded that acting on climate change is both good and necessary:

> The capacity to deliver a plan, such as this, which reaches so far into our daily lives, will be impossible without the support and acceptance of the value of the changes being targeted and the necessity of the

> key infrastructures which support them. (Department of Communications, Climate Action and Environment 2019: 12-13)

A 'key infrastructure' is never mentioned anywhere in the document. Language. There is no Department of 'Communications' without language. No Climate Action Plan can be written or understood without – language. Elsewhere, we are told that 'every network which supports our lives […] must adapt rapidly' and the networks are listed off, 'energy, transport, telecommunication, public service, waste management' (8). Again, there is no mention of language. Try running a complex society without the symbolic resources of language, but obviously the most important network of all in a moment of severe crisis can be serenely ignored. More tellingly for the purposes of this work, in the 148 closely written pages of the Climate Action Plan, there is not one single mention of the Irish language. The language that has described, mapped and narrated for over 2000 years the island nation that the Plan purports to save is completely absent. In the tens of thousands of words in English, not a single word about Irish.

In a section entitled 'The Critical Nature of the Challenge' the authors of the Plan propose 'a big step-up in our engagement with citizens and communities through more coherent mobilisation of existing structures and initiatives to inform, engage, motivate and empower people to take climate action' (17). Where is the coherence in the failure to mobilise the language that has for centuries been central to the Irish engagement with the natural environment? How are citizens to be engaged, motivated and empowered if one of the most precious resources they have for understanding the place of their island in the world of earth, sky and sea is routinely and systematically ignored? The wholesale capture of public discourse on climate change by the baggy, technocratic abstractions of 'structures and initiatives' means that 'citizens and communities' are ultimately treated as wordless creatures responding in Pavlovian fashion to the fiscal versions of carrot (grants) and stick (taxes). This severely reduced notion of human agency does not bode well for the future effectiveness of plans to transition to a carbon-neutral society. Indeed, it has been the very neglect of the historical and rhetorical dimensions to human actions and beliefs that has so powerfully enabled the climate scepticism of populism to thrive across the globe. When no credence is given to the mythical power of language, myth-making of the most toxic kind will thrive.

When the *Climate Action Plan 2019* tells its readers that 'There is an onus on each of us to mitigate the magnitude of long-term climate change by taking actions to reduce greenhouse gas emissions' (15), one wonders how citizens are supposed to develop an ability to cultivate 'long-term' thinking when they are explicitly deprived in the Action Plan of the key to a long-term vision of their island home, namely, the Irish language. In thinking about the Irish language and climate change we might begin by considering the long-term fallout of language shift in Ireland and how this contributes to blind spots which affect our current ecological predicament.

Patricia Palmer, David J. Baker and Willy Maley in their presentation of a research project on Early Modern Ireland sketch out the perverse utopia of the colony:

> The colonial imperative is to destroy all memory of what went before; the new order still founded on amnesia. The defeat of the Irish Confederacy at the end of the Nine Years War reduced the country, in the eyes of the victors, to a clean slate, a *tabula rasa*. Fynes Morryson, Lord Deputy Mountjoy's secretary, declared that, after the Flight of the Earls, "Ireland was left as a payre of cleane tables, where in the state might write lawes at pleasure". (Palmer, Baker and Maley 2019)

Central to the project of erasure is language. In the oft-quoted line from Edmund Spenser, 'the speech being Irish, the heart must needs be Irish for out of the abundance of the heart the tongue speketh' (Spenser 1970: 68). As language situates people in their environment in terms of both description and narration – telling you where you are and what's around you and where you come from – the project of removing Irish from public life has as one long-term consequence the alienation of people from their own surroundings. The schoolmaster Hugh in Brian Friel's *Translations* cautions 'that words are signals, counters. They are not immortal. And it can happen – to use an image you'll understand – it can happen that a civilisation can be imprisoned in a linguistic contour which no longer matches the landscape of... fact. (Friel 1981: 43). But conversely, it can also happen that a people can find themselves imprisoned in a linguistic contour which no longer matches the landscape *tout court*. The population shifts to a language which bears no relationship to the environment in which they find themselves. The ecological consequences are profound in that the connection to place and history – a sense of which is central to the creation of sustainable and resilient localities – is seriously fractured. As Palmer and her colleagues note, 'We may imagine ourselves at an angle to the Anglosphere, basking in our guilt-free positioning as both recovering colony and third richest country in Europe, but we have little countervailing sense of what exactly the absence that haunts our modernity might be.' (Palmer *et al* 2019).

Historians such as Vincent Morley and Nicholas M. Wolf have rightly pointed to the resistance put up to the attempted removal of Irish from public life in Ireland, but ultimately the colonial project of language substitution was strikingly successful (Morley 2017; Wolf 2014). It is the absences from this past that are now coming back to haunt our present in terms of our relationship to the environment. Two months after the publication of the Climate Action Plan and four months after Dáil Éireann declared a climate emergency, the Republic of Ireland was confirmed in The Environmental Indicators Ireland report in August 2019 as having the third highest emissions of greenhouse gases per capita in the EU. Ireland generated 13.3 tonnes per capita of CO_2 equivalent in 2017, and was only exceeded by Estonia and Luxembourg in that year. This figure was more than double the rate in Sweden, which had the lowest rate of emissions in 2017 at 5.5 tonnes per capita (Central Statistics Office 2019). If Ireland consistently comes close to bottom of the European class in terms of environmental performance does this not indicate a deeper malaise in terms of its citizens' relationship to the environment? Is there a sense in which a reckless disregard for biodiversity, painstakingly documented by Pádraig Fogarty in his work, *Whittled Away: Ireland's Vanishing Nature* (2017), is not born out of a disengagement from an environment whose human and natural history is secreted away in a language that receded from common reach in post-Famine Ireland? Is the triumph of a culture of possession over a culture of dwelling – dramatically illustrated by the boom–bust cycles of Irish property markets – not an illustration of the divorce of land from sense, the having being everything and the understanding (of context, history, sustainability) counting for nought?

When Fynes Morryson wrote, "'Ireland was left as a payre of cleane tables, where in the state might write lawes at pleasure', he spells out the extractivist fantasy that drove imperial, and later corporate, scrambles for resources and fossil fuels. Imagine the conquered land or the acquired property as *terra nullius*, to be disposed of in such a way as to yield the maximum, short-term profit. In this sense, the Tudor experiment in (language) extinction and (territorial) extraction makes Ireland the ideal laboratory for a form of ecological dispossession that will be replayed endlessly in various corners of the Empire. One consequence is the need for the *re-indigenisation* of Irish language politics in the age of the Anthropocene. To suggest what this might mean I want to look briefly at the writings of a linguist and a political activist.

Indigenisation

Arran Stibbe, towards the end of his work on the emerging discipline of ecolinguistics makes the following observation:

> There is, however, a shortage of voices, in this book and in ecolinguistics in general, from traditional and indigenous cultures around the world. Within these cultures are a great multitude of stories, some of which may be invaluable in the reinvention of self and society in the transition to new ways of living and being. (Stibbe 2015: 193)

The importance of 'traditional and indigenous cultures' is not simply to do with past stories as a source of future change but with present stories as a pointer to imminent survival. Naomi Klein notes that running economies on energy sources which release poisons into the atmosphere as an inescapable part of their extraction and refining has always necessitated the existence of 'sacrifice zones.' These zones have a number of features in common:

> They were poor places. Out-of-the-way places. Places where residents lacked political power, usually having to do with

> some combination of race, language, and class. And the people who lived in these condemned places knew they had been written off. (Klein 2014: 310).

Gaeltacht areas in Ireland often functioned as our own sacrifice zones, less because they had to play handmaid to extractivist industries, with the exception of the Shell pipeline in Rossport, Co. Mayo, and more because they were places 'where residents lacked political power, usually having to do with some combination of race, language, and class'. It is often the poorest people on the planet speaking lesser-used languages in more remote parts of the world that find themselves at the frontline of the race to extract as much fossil fuel resources as possible from the earth. Describing how the Beaver Lake Cree Nation took on oil and gas companies over tar-sands oil extraction, Klein claims that 'some of the most marginalized people in my country are [...] taking on some of the most wealthiest and most powerful people on the planet' (379). The move to a carbon credits system, a market solution that allows wealthier countries and corporations to offset carbon emissions by buying carbon 'credits' from countries with low emissions, has, indeed, made the situation of indigenous peoples even more precarious.

Bram Brüscher, a geographer, has coined the term

'liquid nature' to describe the way in which fields, forests and mountains lose their intrinsic, place-based meaning and become deracinated, abstract commodities in a global trading system (Brüscher 2013: 20-36). In the Bajo Aguán region of Honduras, owners of palm-oil plantations have been able to register a carbon-offset project that they claim captures methane. Encouraged by the promise of financial gain from captured gas, huge tree farms have disrupted and displaced local agriculture. A resulting violent cycle of evictions and land occupations had by 2013 led to the deaths of up to a hundred local farmers and their advocates (Wong 2013). As Klein remarks, 'it's easier to cordon off a forest inhabited by politically weak people in a poor country than to stop politically powerful corporate emitters in rich countries' (Klein 2014: 223). An additional irony is that it is indigenous peoples practising some of the most sustainable, low-carbon lifestyles on the planet who are being sacrificed for the carbon market. If narratives around Irish have understandably focused on political history and the power dynamics of the Anglo-Irish relationship, it is crucial in the light of what Stibbe and Klein have to say about indigenous peoples that we now globalise rather than nationalise the Irish language. In other words, it is by situating the Irish language in the context of indigenous ecological struggles around the world that we end up with a more realistic and more fruitful concept of what is necessary and what is possible in terms of relating indigenous language rights to environmental progress.

This formation of these new, transnational alliances will have to deal with a repeated misapprehension about indigenous cultures. The standard trope for the protection and promotion of indigenous languages is salvage. A culture or language must be retrieved lest it disappear forever. The focus is always on a legacy in peril. Mark Abley in *Spoken Here: Travels Among Threatened Languages* speaks of the Kimberly region in Western Australia and notes that of the thirty different languages only three are spoken by children: 'It's as though twenty-seven sphinxes are crumbling away before our eyes' (Abley 2003: 35). The image is that of the sphinx and a remote past. Similarly, from its roots in 18th- and 19th-century antiquarianism, a great deal of the language around 'saving' the Irish language has focused on its former glories, its antiquity, the immeasurable richness of what Douglas Hyde described as 'the nation which was once, as everyone admits, one of the most classically learned and cultured nations in Europe' (Hyde 2019 [1892]). The problem with this salvage archaeology is that it condemns indigenous cultures to the quarantine stations of the past. Daniel O'Connell, for example, was perfectly content to associate loyalty to Irish as sentimental whimsy which had no practical purchase on the Irish present or future:

> I am sufficiently utilitarian not to regret its [the Irish language's] gradual abandonment. A diversity of tongues is no benefit; it was first imposed upon mankind as a curse, at the building of Babel. It would be of great advantage to mankind if all the inhabitants of the Earth spoke the same language. Therefore though the Irish language is connected with many recollections that twine around the hearts of Irishmen, yet the superior utility of the English tongue, as the medium of all modern communication, is so great that I can witness without a sigh the gradual disuse of Irish. (Daunt 1848: 14-15)

It is precisely this false timeline which is challenged by contemporary students of indigenous communities across the globe. It is indigenous groups living in sacrifice zones, however, who have an understanding of the relationship of the human to the non-human world which becomes especially valuable in an era of environmental degradation.

Elizabeth Povinelli has coined the term 'geo-ontology' to speak about the worldview of many indigenous groups in Australia who do not establish rigid conceptual and hierarchical distinctions between the human and the non-human. Treating rock formations as inert, inanimate matter does not make sense from their geo-ontological perspective where the world in all its variousness is an active part of an animate cosmos (Povinelli 2016). The political theorist Jane Bennett in *Vibrant Matter: a political ecology of things* argues for a 'vital materiality' that runs through and across bodies, both human and non-human. She claims that the 'quarantines of matter and life encourage us to ignore the vitality of matter and the lively powers of material formation' (Bennett 2010: vii). The point is not an appropriative exoticism which sees endangered languages as the receptacles of past greatness or vanished worlds but an argument which construes these languages and cultures as part of future-oriented cosmologies (Danowski and Viveiros de Castro 2014: 221-339; Latour 2017). Even the most cursory examination of Irish folklore or mythology points to this repeated and systematic engagement with the non-human world. As Dáithí Ó hÓgáin pointed out in *Myth, Legend and Romance: An Encyclopaedia of the Irish Folk Tradition*, 'People in Ireland [...] have speculated on the nature of the physical world around them, giving rise to a rich, imaginative lore' (Ó hÓgáin 1990: 177). Describing the prevalence of animal motifs in Irish folklore Ó hÓgáin argues, 'The various skills and characteristics of fauna have long provided inspiration to poets and storytellers. In ancient times [...] seers were believed to go about in the form of animals and thereby gain knowledge from other than human perspectives' (31). It is precisely this capacity to begin to view the planet from 'other than human perspectives' which is such a necessary skill in dealing with imminent climate catastrophe.

What constitutes the richness of these perspectives is that they provide us not with a way of celebrating the past but of constructing the future. Rather than being seen as incontrovertible evidence of the credulous naiveté of the 'primitive' mind, they offer a way of engaging with the non-human world which can teach us invaluable skills for a future where the non-human is coming back to exact climate revenge on human hubris. In other words, as the French theorist Bruno Latour has repeatedly observed, the supposed pragmatists, the Connellite utilitarians who believe in the supreme value of monocultures and monolingualism, are the hopeless idealists. They remain wedded to a fantasy of an instrumental language spoken by an all-powerful subject who can go on exploiting the non-human world indefinitely. As Latour points out in *Où atterrir: Comment s'orienter en politique* (2017):

> Au bilan la seule chose qui compte, ce n'est pas de savoir si vous êtes pour ou contre la mondialisation, pour ou contre le local, c'est de comprendre si vous parvenez à enregistrer, à maintenir, à chérir le plus grand nombre d'alternatives à l'appartenance au monde. [In the end the only thing that matters is not whether you are for or against globalisation, for or against the local, but how to record, maintain and cherish the greatest possible number of ways of belonging to the world] (Latour 2017: 26-27)

The future lies with languages that can maximise diversity, alter human interaction with the non-human and reveal ways of being that are connected to the specificities of place but are open to the world. Irish is one of those languages of the future. The genuine pragmatists are those who look to the language for resources to build an alternative future, not the monolingual fantasists who would condemn us to an intolerable past in the name of an unsustainable future.

It is in the context of the culture of extractivism, sacrifice zones and the liquefaction of nature through the creation of carbon markets, that a future role for the Irish language becomes manifest. If 'language' is in there with 'race' and 'class' as a marker of peripherality and disadvantage, then struggles around language advocacy and translation rights become part of the broader ecological struggle. Russ Rymer, in an article discussing the predicament of endangered languages, argues that:

> Small languages, more than large ones, provide keys to unlock the secrets of nature, because their speakers tend to live in proximity to the animals and plants around them, and their talk reflects the distinctions they observe. When small communities abandon their languages and switch to English or Spanish, there is a massive disruption in the transfer of traditional knowledge across generations – about medicinal plants, food cultivation, irrigation techniques, navigation systems, seasonal calendars. (Rymer 2012)

It is striking that the 'traditional knowledge' Rymer lists has primarily to do with ecological knowledge, the things that a community needs to know in order to maintain a sustainable lifestyle. Cmiique Iitom is a language spoken by the Seri people of Mexico, traditionally seminomadic hunter-gatherers living in the western Sonoran Desert near the Gulf of

California. Cmiique Iitom has terms for more than three hundred desert plants. The Seri word for harvesting eelgrass offered a clue to scientists about its nutritional merits and it turned out that its protein content was comparable to that of wheat. Seri names for animals reveal behaviour that were treated with initial scepticism by scientists:

> The Seris call one sea turtle *moosni hant coit*, or green turtle that descends, for its habit of hibernating on the floor of the sea, where the traditional fishermen used to harpoon it. "We were sceptical when we first learned from the Seri Indians of Sonora, Mexico, that some Chelonia are partially buried on the sea floor during the colder months", stated a 1976 paper in *Science* documenting the behaviour. "However, the Seri have proved to be highly reliable informants". (ibid)

The case for the value of Cmiique Iitom is articulated at two levels, one case involving impossibility, the other, possibility. At one level, the language is valuable because it contains words or terms for which there is no equivalent in other, particularly dominant languages, an example being the many unique terms for desert plants. Zero-equivalence here is a version of the 'untranslatable' as there is nothing to translate the terms into (though this does not mean that they cannot be translated). They only exist in the Cmiique Iitom language. Similarly, a lack of a term can point to the specificity of a culture rooted in a particular ecological niche. Rymer claims that in the Aka language, spoken in Palizi in Arunachal Pradesh, India's most northeastern state, there is no word for a job in the sense of salaried labour.

In the 2015 general election manifesto for the British Conservative Party under the heading, 'We will promote integration and the British values', voters were told that:

> Being able to speak English is a fundamental part of integrating into our society. We have introduced tough new language tests for migrants and ensured councils reduce spending on translation services. Next, we will legislate to ensure that every public sector worker operating in a customer-facing role must speak fluent English. And to encourage better integration into our society, we will also require those coming to Britain on a family visa with only basic English to become more fluent over time, with new language tests for those seeking a visa extension. (Conservatives 2015: 31)

The 'tough new language tests' and the reduction of 'spending on translation services' make language the litmus test of integration. Assimilation is mastery of the dominant host language and a refusal of the translation process. The language ideology informing these commitments is firmly rooted in a notion of monolingualism that is historically a relatively recent phenomenon in world history. Yasemin Yildiz has argued, emerging in the age of nations, it is 'monolingualism, not multilingualism, that is the result of a relatively recent, albeit highly successful, development' (Yildiz 2012: 3). Monolingualism becomes much more than a quantitative term to isolate and designate a particular language but emerges as 'a key structuring principle that organizes the entire range of modern social life from the construction of individuals and their proper subjectivities to the formation of disciplines and institutions, as well as imagined collectives such as cultures and nations' (3).

A political ecology of language needs to view languages in their connectedness, not in their isolation. That is to say, a language can be seen as offering its speakers a set of possibilities in terms of understanding or relating to the world, what is often called 'culture', but it cannot be reduced to that. As Heller and Duchêne have pointed out the equation of one language with one culture simply maintains, 'the language-culture-nation ideological nexus' (Heller and Duchêne 2007: 7). Historians of translation have shown that languages are endlessly open-ended, repeatedly subject to the influences of other cultures and languages even if translation has also served to define and maintain the contours of language (Delisle and Woodsworth 2012). An ecological championing of language diversity cannot be a simple, mimetic reflection of the monolingual language ideology which underpins, for example, the defence of 'integration and British values'. Rather, in terms of a globalisation from below, an ecological case for Irish must be alert to those practices which result in contact across a wide range of languages and cultures in a plurality of urban and rural settings and where the complex history of languages trouble any easy symmetry between mono-language and mono-culture.

If ecology is first and foremost about establishing the conditions for the continued possibility of life, then there is a further lesson to be learnt from indigenous peoples across the globe. Donal Casey in a short article published in IrishCentral.com opens with a statement that is anything but controversial, 'The debate about whether the Irish Gaelic (Gaeilge) language is dead or alive still remains a hot topic in Ireland' (Casey 2018). Living speakers of Irish often take this as a personal affront, a form of denial specific to Ireland, but the pattern of erasure, the lowering of language into the grave of historical oblivion and political indifference, is repeated in many locations. On the 28 December 1890 one hundred and fifty men, women and children of the Lakota tribe were brutally massacred on the Pine Ridge Reservation near the border of South Dakota and Wyoming. What became known as the Wounded Knee massacre was portrayed as the terminal point of native American culture. Dee Brown's *Bury My Heart at Wounded Knee* (1970) gave popular credence to this myth of the end of a civilisation. As Ari Kelman has observed, after Wounded Knee there was believed to be nothing but eternal silence, 'It is as though Native America ceased to exist after 1890; a rigid chronology somehow became a bleak fate that eventually ossified into an inescapable teleology' (Kelman 2019: 5). There is no room for the stories of the Native nations that survived and thrived.

David Treuer in *The Heartbeat of Wounded Knee* claims that the accepted truth after the Lakota massacre became, 'There had been an Indian past, and overnight, there lay ahead only an American future' (Treuer 2019: 6). In the Irish version of this 'rigid chronology', Kinsale, the Flight of the Earls and the Great Famine become the staging posts for the vanishing Gael. Before, there had been an Irish-speaking past, there lies ahead only an English-speaking future. In the Manifest Destiny of monolingualism the presence of living communities and literatures and broadcast media in indigenous languages is an anachronism, a historical anomaly that invites either satire or amnesia. Treuer argues that the most enduring tragedy for the Native American is the opacity of the label, 'tiny, fretful intricate details are what make us who we are. And they are lost again and again when we paint them over with the tragedy of "the Indian"'. And he adds that, 'in this sense, the victims of Wounded Knee died twice, once at the end of a gun, again at the end of a pen' (170). If many Irish speakers did indeed die during the wars of conquest at the end of a gun, they have experienced a second death at the end of the pen of dismissive indifference. One example among many, a television columnist writing in *The Irish Times* claimed, 'most of us can hardly remember more than a cúpla focal of what was so badly taught at school. And we couldn't be bothered anyway' (Harrison 2012: 8). What indigenous comparisons show is that there nothing anomalous about the situation in Ireland. Being mindful of this can then both enlighten and embolden those who seek a more linguistically and ecologically viable way of dwelling in the world. Not as ghosts but as actors.

Thinking Outdoors

If many indigenous cosmologies are primarily focused on the world outdoors (Salmond 2017), the history of industrialised modernity is largely about staying indoors. As the French writer, farmer and activist, Pierre Rabhi, has pointed out, confinement is the defining experience of modernity. Human beings, he argues, in developed countries, 'de la maternelle jusqu'à l'université, ils vivent un enfermement' [from pre-school to university they are kept inside] (Rahbi 2010: 41). Afterwards, they go to work in government buildings or office blocks. They leave these buildings to get into cars that prolong this isolation from the world beyond the (wind)screen. At night they look out at that world from behind another screen – television, laptop, tablet. As the essayist Chris Arthur has pointed out, the result is that 'millions of us have become more defined by the time we spend indoors than outside. As a result, we've lost touch with the natural world. Those with the most pronounced indoor sensibilities can't identify even the commonest species of birds, insects and plants in their locality' (Arthur 2019: 60). Thinking outdoors is not a skill cultivated by our educational system. The primary school curriculum makes a principled effort to engage with the environment but results are fitful at best, and by the time children are herded into the exam factory farms of the post-primary sector, contact with the great outdoors is a distant memory of 'nature walks' in Senior Infants.

Chris Arthur asks the question, why is it, 'that a garden, or small wooded area, a field, a stream, or patch of marshland are not also thought essential adjuncts for the business of learning. Is it any wonder children are ill-attuned to nature if their education happens at arm's length from it?' (61). The need for outdoors thinking is crucial in the case of Irish. Where does most instruction in Irish take place? Indoors. What is most people's memory of learning Irish? Being indoors. There is the welcome parenthesis for many schoolchildren of the Gaeltacht summer colleges where part of the pleasure is learning the language outside. But it remains that, a brief parenthesis, bracketed by years of classroom confinement. So, we have the rich irony of pupils studying texts by poets and prose writers on a natural world they never get to experience. In the understandable desire to be taken seriously as speakers of a modern language fit for purpose in urban spaces, Irish-language speakers have been eager to shake off sentimentalised associations with the Wild Atlantic Ways of rural kitsch – setting suns, thatched roofs, beached currachs. There is a need now, however, to re-engage with the natural environment before it falls victim to the forms of indoor thinking that blind us to our vital connection with the living, organic reality of the island. The philosopher Peter Carruthers writing about the ethics of land use argues that philosophy and ethics are formed by 'real-life experiences' and that thought is deeply rooted 'in experiential reality and authentic encounter' (Carruthers 2009: 310). He claims:

> We need, then, to 'enhance our land experience' and to create opportunities for people (perhaps especially young people and city-dwellers) to encounter land, nature and farming, both directly and through art, literature and music. These encounters can stimulate imagination, create vision and build a kind of 'moral credit' (310)

The Irish language has ample resources in terms of stories, terminology, academic studies, online materials to underpin this urgent enterprise of taking Irish thinking outdoors. A living language for a living planet is infinitely preferable to a dead language on a dead planet. Leaving Irish to languish indoors in the classroom robs it of its relevance to understanding the ecological realities, past and present, that will irrevocably shape Irish futures. Chris Arthur warns of the dangers of escaping indoors:

> If more of us spent more time outside, directly engaged with nature, it would add

> the iron of firsthand experience to what may otherwise be lukewarm conviction based on something that's too abstract to engage us. Our tragedy as a species may be by the time the impact of climate change reaches into even the most complacent indoor bastion of the psyche, it may be too late to do anything about it (Arthur 2019: 64).

Spending 'time outside' is not, however, likely to be very profitable if we do not know what we are looking at. The Indo-European root of the Irish word for poet, *file*, relates to seeing, so a crucial contribution of Irish is to allow us to see better, more deeply, in a more context-sensitive way.

A failure to see means a failure to see a great many things and not just the landscape before you. The English nature writer, Robert Macfarlane, describing his reactions to Barry Lopez's travel book, *Arctic Dreams*, claims that one of the lessons the book taught him was that, 'while writing about landscape often begins in the aesthetic it ends in the ethical.' Lopez's intense attentiveness was 'a form of moral gaze born of his belief that if we attend more closely to something then we are less likely to act more selfishly towards it' (Macfarlane 2015: 211). In other words, attending to language is a matter not of nostalgia but of urgency.

The American poet and farmer Wendell Berry sees language as a way forward, not a step backwards:

> People *exploit* what they have merely concluded to be of value, but they *defend* what they love and to defend what we love we need a particularizing language, for we love what we particularly know. (Berry 2000: 41)

How this plays out politically can be illustrated by an example from Scots Gaelic. The engineering company AMEC supported by British Energy submitted a proposal for the development of a wind farm on Brindled Moor on the Isle of Lewis in the Outer Hebrides. The proposal consisted of a wind 'power station' of 234 turbines each 140 metres high with a blade span of 80 metres (the terminology of 'farm' tries to mask the reality: these installations are dispersed power stations). Each turbine required a foundation of 700 cubic metres of concrete and '5 million cubic metres of rock and 2.5 million cubic metres of peat would be excavated and displaced' (Macfarlane 2015: 28). One response was the publication of a booklet *A-mach an Gleann/A Known Wilderness* where the co-authors, Anne Campbell and Jon McLeod mapped their moor-walks, describing in great detail features of landscape, wildlife, folk memory that they observed or remembered on the way.

The second was the compilation by Anne Campbell and Finlay McLeod of *Some Lewis Moorland Terms: A Peat Glossary*, a glossary of 126 Gaelic terms to describe features of the moorland such as *teine biorach*, 'the flame or the will-o'-the-wisp that runs on the top of heather when the moor is burnt during the summer' or a *rùdhan*, 'set of four peat blocks leaned up against one another such that the wind helps their drying' (18). What the *Peat Glossary* and *A Known Wilderness* are contesting is the notion of the moorland as a *terra nullius*, ripe for development and extraction. When James Carnegy-Arbuthnot, an estate owner in Angus, argued in 2013 that it was right that so few people own most of the land in Scotland because most of it was 'unproductive wilderness', he was readily equating emptiness of landscape and plenitude of possession. What the excluded language does is to restore a sense of history, particularity, complexity and (common) ownership to place. Noteworthy in the responses to the AMEC development is that outdoor thinking and language – the moor-walks and the glossary – are both intertwined in opposition to the proposed project. The detailed and loving description of place becomes inescapably ethical. In a broader sense, what the strategic minoritisation of the major languages of industrial development involves is an overall shift from an ideology of extractivism to an ideology of regeneration. This is a regenerative move that challenges through the energies of the minoritised language the predatory emptying out of landscape and language.

Internalisation

The *Climate Action Plan 2019* is blunt about future challenges. There is no doubt but that the changes resulting from global warming will 'cause extensive direct and indirect harm to Ireland and its people' (Department of Communications, Climate Action and Environment 2019: 15). However, on a more positive note, 'There is now a higher level of awareness and understanding in Ireland of the impacts of climate change' (15). This is unquestionably true but what about the consequences of those consequences? Are we ready to think through the radical implications of the changes necessary in the new climate regime? One such change is the coming revolution of internalisation. Paul Virilio, for example, points out in the recent history of globalisation:

> L'importance accordée aujourd'hui, par les enterprises, à l'EXTERNALISATION de leur production – y compris celle de la recherché et du développement de l'innovation – au detriment de leur ancienne localisation patrimoniale. Ainsi, depuis plusieurs années, l'*extérieur* l'emporte partout sur l'*intérieur* [the importance [...] attached by firms to the EXTERNALISATION of their

> production – including research and development – at the expense of their former specific local base. So, for several years now, the *external* has become more important everywhere than the *internal*] (his emphasis] (Virilio 2009: 19).

A graphic illustration of the practice of externalisation is the purchase by Daewood Logistics of 1.3 million hectares of arable land on the island of Madagascar. The purpose is to provide a secure food source for the population of South Korea but 70% of the island's population currently live below the poverty level (66). What is more, the island has been identified by the biologist and zoologist Edward Wilson as one of the planet's twenty-five 'biodiversity hotspots'. These hotspots only cover 1.4% of the earth's surface but account for 43.8% of all known plant types and 35.6% of species of mammals, birds, reptiles and amphibians living on earth (Wilson 2003). Externalisation of course is a core feature of the post-Fordist economy and the very term 'globalisation' indicates the global reach of the external, where the British working class are now said to live in China (source of so many manufactured goods) and where what the Indian scholar Harish Trivedi has termed 'cybercoolies' in low-cost call centres service wealthy customer bases in advanced industrial nations (Trivedi 2004). In a sense, forms of externalisation have been in evidence ever since David Ricardo formulated his theory of comparative advantage where industrialised nations import agricultural produce from nations that enjoy the comparative advantage of still possessing fertile land for the production of food (Ricardo 2004). These nations in turn buy manufactured goods with earnings from the sale of foodstuffs and begin their own process of industrialisation.

The difficulty is that the form of intensive food production and population growth which exhausts the soils of industrialised nations eventually affects those nations enjoying a comparative advantage in the realm of agriculture, and they too face the problem of finite natural resources. The limits to externalisation are evident in what we now know about human impact on terrestrial ecosystems. Between 40% and 50% of the earth's surface has been modified by human activity, the concentration of carbon dioxide in the earth's atmosphere has increased by around 30% since the beginning of the industrial revolution, half of the world's water reserves are used by human beings, in recent history a quarter of the world's species of birds have been wiped out and two thirds of the world's fish species are currently fully or overexploited (Intergovernmental Science-Policy Platform on Biodiversity and Ecosystem Services 2019).

If the planet is to survive, the revolution of

externalisation which marked the modern era must give way to a revolution of internalisation in the era of human-induced climate change. In other words, the areas of transport, industry, architecture, agriculture, energy, all will have to move away from the compulsive externalisation of the globalised market economy and move to practices of greater internal self-reliance and resilience. In describing a future revolution of internalisation we need to be clear about the implications for a progressive view of language and society.

In the kinetic euphoria that accompanies contemporary eulogies to the global, the dizzying rate of global exchange is often equated to the relentless onward march of liberal democracy. The defensive local allegiance to the particular has no place in the fraternal embrace of the universal. Enlightened supranational governance or the inclusive joys of global citizenship are the termini of the accelerated voyage away from the retrograde claims of the specific. European and American populism with their aggressive defence of national interest seem to be cautionary tales in this respect. For progressives, universal love for all peoples and all species represents a kind of radical externalisation of sentiment which envelops the planet in an outward burst of fellow-feeling. However, are local/internal repudiation and global/external celebration the only grounds for intelligent sympathy or due care?

More than twenty-five years ago, the philosopher and critic Val Plumwood expressed a useful scepticism with respect to the viability of sympathies which are too indiscriminate in their focus:

> This "transpersonal" identification is so indiscriminate and intent on denying particular meanings, it cannot allow for the deep and highly particularistic attachment to place that has motivated both the passion of many modern conservationists and the love of many indigenous peoples for their land. (Plumwood 1994: 152)

The danger with loving everything is you end up loving nothing. Love that is indiscriminate leads ultimately to a form of indifference as the object of love is simply one undifferentiated mass. Plumwood argues that, in fact, it is internalisation not externalisation which becomes the effective basis for a meaningful engagement with struggles and issues at home and elsewhere:

> Care and responsibility for particular animals, trees, and rivers that are known well, loved and appropriately connected to the self are an important basis for acquiring a wider, more generalized concern. (145-6)

It is in fact the local connection, the ready identification of a particular animal, tree, river, predatory oil company or abusive state practice in a local setting which makes possible the ethical imagining of the importance of species preservation or social justice in other places across the globe.

In order to do this, however, citizens need access to the linguistic resources which allow for knowing, loving and connecting to particular animals, trees, and rivers. In Ireland, these resources exist in abundance in the Irish language. Internalisation becomes a highly effective form of mobilisation in starting from concrete, near-to-hand examples to address issues that have a resonance beyond a specific locale. The movement inwards is an opening up, not a shutting down. As Alastair McIntosh has observed, 'I must start where I stand. As children, we used to be told that if you dug a really deep hole, you'd come out in Australia. I think in some ways this is very true. If any of us dig deep enough where we stand, we will find ourselves connected to all parts of the world' (McIntosh 2002: 7). We can only start to dig if we have the appropriate tools. The knowledge and use of Irish is what allows us to dig deep, to understand where we start from. In this way, we will 'find ourselves connected to other parts of the world', not least in a proper appreciation of the vitality of local knowledge for a resilient future.

In the cartoon history of Irish modernity, the setting is a familiar one. The people of Ireland are sunk in the deep sleep of protectionism, lulled into in a trance of self-abnegation by the Gaelic sermonising of their priestly betters. Along come Whitaker and Lemass, the Titans of Progress. The country is saved in the nick of time from disappearing down the plughole of national self-righteousness and the course is firmly set for redemption through free trade (Garvin 2005). In these endlessly rehearsed versions of how we went from preventing the future to promoting it, the Irish language is corralled in with land and religion as a shibboleth of the anti-modern, a form of totem worship for recalcitrant hobgoblins wedded to a vision of frugal autarky on the Western seaboard. Missing from these accounts is what might be termed the strain of "Gaelic modernity" in Irish life, economy and society. This is a form of modernity that calls into question the pat oppositions that even to this day structure debates around language and identity in Ireland. It is also a form of modernity that anticipated much of contemporary ecological thinking. An example of this proto-ecological modernity which we will consider in more detail are the activities of the Irish-language organisation, Gael Linn (Ní Chinnéide 2013).

The writer and language activist Máirtín Ó Cadhain was particularly preoccupied with the economic plight of the Gaeltacht. He was withering on the urban *gaeilgeoirí* who rhapsodised on the virtues of the Gaelic way of life while condemning the surviving speakers of the language to a life of unrelenting poverty at home or permanent exile abroad. Throughout the 1950s and 1960s Gael Linn, an organisation founded by former members of a student Irish-language association in University College Dublin, responded to these concerns by financing a series of initiatives in fishing, metalwork, market gardening and animal husbandry with the express aim of providing livelihoods for people living in Gaeltacht areas. Not all of these projects met with success. One particularly unhappy scheme involved raising rabbits for export to markets on the continent. In 1961, hurricane Debbie swept through Connemara taking in its wake the roof of the large shed in which the animals were kept. The rabbits were not slow in seizing the opportunity to head for the hills and it was eventually decided that *lapin* would remain off the Gael Linn menu. Underlying these initiatives, however, was a welcome realism about the link between language and livelihood which would eventually lead to the establishment of Gaeltarra Éireann and Údarás na Gaeltachta as development authorities in Gaeltacht areas.

More broadly, in terms of the concept of Gaelic modernity, what the Gaeltacht projects pointed to was a desire to incorporate modernity in a framework of local sustainability. The intellectual impetus for this came in part from the members of the Comhchaidreamh, the UCD student body, who worked in semi-state bodies like Aer Lingus, CIE, Comhlucht Siúicre Éireann (the sugar company) and the ESB and who envisaged a fruitful alliance between science, technology, business and indigenous development. The establishment of Inisfree Handknits, the furniture brand Crannac, the trading company Irlandia and the investment fund Cara-Chumann, as well as other initiatives, were part of a philosophy of economic development that would draw on place and cultural specificity to create employment that was local, innovative and attuned to shifts in global demand. The perennial difficulty for these proto-ecological projects was scale. While jobs were created and income was generated, the quantity or amount were scarcely sufficient to meet the demands of an expanding population.

Another example of the connection between language and eco-cultural sensitivity was the development of the music branch of Gael Linn. One of the immediate difficulties Gael Linn faced with its weekly sponsored programme on national radio was not so much what to say as what to play. Despite the activities of recording companies in the United States in the earlier part of the century, there was a dearth of recordings of traditional Irish music and song. Under the label Ceolta Éireann, over twenty recordings were issued with particular emphasis on Irish-language song. Contrary to modernist misconceptions about a land beholden to the theme music of the crossroads, for much of the 1920s, 1930s and 1940s, in Irish broadcasting circles and beyond, there was a widespread petit-bourgeois disdain for the music of the common people. This would change with the pioneering work of Gael Linn and the visionary presence of Riobard Mac Góráin. It was Gael Linn who would provide Seán Ó Riada and Ceoltóirí Chualann with both the financial support and the public exposure that would allow them to pursue their quiet revolution in music in the period, culminating in the legendary concert in the Gaiety Theatre in 1969. The recording from that concert, Ó Riada *sa Gaiety*, is still one of the bestselling records in the Gael Linn catalogue. The involvement with a classically trained composer was indeed part of a broader attempt to bring the folk and classical traditions into a musical conversation that would go beyond the sniffiness and suspicion that had largely dogged the relationship between them up to this point. It was in the Gaiety Theatre, ten years earlier, on March 15th, 1959, that Gael Linn brought together the RTÉ Light Orchestra, the classical pianist Charles Lynch, the Belfast fiddler Seán Maguire, the sean-nós singer Seán 'ac Dhonncha and others in an effort to end the performance apartheid that bedevilled the Irish musical scene. 1967 Gael Linn issued a record, *Ceol na nUasal* (Music of the Nobility), which showcased the work of Turlough O'Carolan and others who had worked between the two traditions. As traditions are only as vital as their renewal, a decision was made to start the Slógadh competition in 1970, a talent competition for young people working with Irish music or song, inspired by a similar competition at the Welsh Eisteddford yr Urdd. At the first competition, the audience was mesmerised by a group of musicians from Donegal playing "Lisa". Clannad would be the first in a long list of groups and performers that would achieve fame after first coming to prominence in the Gael Linn competition. Others included Skara Brae, The Hothouse Flowers, Na Casaidigh, Stockton's Wing, Sharon Shannon, Iarla Ó Lionáird, Declan Masterson and Kíla. Critical to the success of Slógadh was the disavowal of the *volkisch* trappings of the more established competitions and the celebration of forms of musical hybridity that proved deeply compelling for the younger musicians working the traditional idiom.

The founders of Gael Linn quickly realised that a culture of complaint was of limited usefulness. Petulance would never amount to a politics. Since independence, a great deal of the energies of Irish language organisations had been directed towards heaping scorn on successive governments for their failure to deliver on commitments to the language. What was lacking were the attempts to develop alternative ways of approaching economic development. In more recent times, economic and business theorists like Finbarr Bradley and James J Kennelly in *Capitalising on Culture, Competing on Difference: Innovation, Learning and a Sense of Place in Globalising Ireland* (2008) have argued that in a post-industrial economy of cognitive and aesthetic goods where comparative economic advantage is embedded in design and informational density, the prime sources of design and informational difference are in the specificities of culture and place. In a world of homogenised goods and services, difference makes a difference as they set out to demonstrate in a series of case studies in *The Irish Edge: How Enterprises Compete on Authenticity and Place* (2013). There is another dimension to the sense of culture and place which is as much rooted in the ecological as in the economic. In this view, the resources of the local become a logical starting point for a response to global vulnerabilities. Tradition becomes an agent of, not an obstacle to, renewal. This is not place as the holy ground of blood and soil irredentism, but place as a fractal opening up to a world of global interdependency. The cartoonists of Irish modernity in pitting Gaelic Reaction against Gallant Reform have prevented different versions of the Irish future from coming into being, versions that have their pedigrees in different, competing forms of Irish modernity.

Guilt

The future tense in a period of rapid climate change is a dangerous tense. What lies ahead may make us regret even more intensely what we have left behind or feel more intensely a regret for what we have refused to leave behind, a form of habitation that is making the earth uninhabitable. In John Lanchester's novel *The Wall*, set in the near future in a Britain adjusting to the effects of severe climate change, Kavanagh, the main character, reflects on the catastrophic legacy of the failure to act on climate change:

> None of us can talk to our parents. By "us" I mean my generation, people born after the Change. You know that thing where you break up with someone and say, It's not you, it's me? This is the opposite. It's not us, it's them. Everyone knows what the problem is. The diagnosis isn't hard – the diagnosis isn't even controversial. It's guilt: mass guilt, generational guilt. The olds feel they irretrievably fucked up the world, then allowed us to be born into it. You know what? It's true. That's exactly what they did. They know it. We know it. Everybody knows it. (2019: 55)

The difficulty is that guilt can be as much a reason for inaction as action. In Ireland, guilt as an emotion which dominated the moral economy of the country for much of the twentieth century was seen as responsible for the general sense of paralysis that Joyce, among others, saw as gripping the island. Concern for an environmental future can shade into forms of ecological moralising where the rising temperatures of the oceans replace the Redemptorist fires of hell. Timothy Morton in *Being Ecological* wonders whether it wouldn't 'be better to stop with the sermonising, the shaming and the guilt that are part and parcel of the theistic approach to life that arose during the agricultural age' (Morton 2018: 35). He argues that, 'Ecological awareness is awareness of unintended consequences. Some ecological politics is about trying to light up everything in a totally nonflickery way, to make sure that there are no unintended consequences.' The only consequence of this would be a 'control society' (50) where everything would be subordinated to a kind of technocratic rationalism that brought about the ecological crisis in the first place. As Morton laments, 'At present the ways in which we talk to ourselves about ecology are stuck in horror mode: disgust, shame, guilt' (60).

Replace the word 'ecology' with 'Irish' and would much in that sentence need to be changed? Is there not much in Irish life to suggest that a major difficulty around the Irish language is not so much a violent, partisan rejection of it – which is now largely a minority sport – but more a sense of 'shame', 'guilt', an unease about or an inability to speak or understand the language. This passive guilt may indeed translate into active hostility as there is no more forceful hatred than self-hatred – hence the ferocity of civil wars. However, it is more often than not the embarrassed confession of inability – generally blamed on the conveniently abstract alibi of the Bad Teacher – that hobbles language revival with the fetters of self-loathing. Put Irish and ecology together and the potential for self-reproach tends towards infinity. Environmental responsibility and language awareness demand our reluctant attention but rarely, it seems, our unprovoked love. Morton claims that this is precisely the problem for ecology, how to disentangle action from coercion:

> Just keeping lifeforms alive is allowing for some kind of pleasure (and don't forget the pleasure of enjoying their existence, and the pleasure of doing less harm). And then I realized that this sort of feeling would be what living in an ecologically attuned society feels like. Instead of policing pleasure we would be inventing new ones. (210)

Thinking ecologically does not mean being hostage to the horror mode. And thinking about Irish does not mean having to indulge in endless sermonising. Connecting to, experiencing, appreciating the world around you in a more intense and meaningful way is a way to suggest new pleasures rather than endlessly castigate old ones.

Fundamental to these new pleasures are new connections. Connections to things you did not know, experiences you have never had, people you have never met. Ecologically, the fundamental connection is to the totality of the world we inhabit, including both the human and the non-human. The malaise or dis-ease that is manifest in climate change is the repudiation of that connection. Overcoming that trauma, discovering the pleasure of reconnection in Ireland is intimately bound up with the rediscovery of the language of that connection, Irish.

We get a glimpse of these new pleasures of reconnection in Tim Robinson's account of climbing Errisbeg Hill near Roundstone in Connemara. He discusses the work of Alexander Nimmo, the engineer who founded the village of Roundstone and who in 1813 carried out a survey of the bogs of west Galway. In the words of Robinson, Nimmo was 'the first to give an overall description of what then was a nameless wilderness' (Robinson 2006: 72). Description was never far from calculation, however. In a submission to a House of Commons commission on the practicality of the cultivation of the bogs in Ireland Nimmo noted:

> The ridges aforesaid prevent any navigation from being easily led into these lakes from the limestone tract [of central Connemara], but in the event of an extended cultivation, something of that kind might be effected; for example, a dam upon the stream near Ballinaboy would throw it into one narrow sheet of water on to the upper end of Lough Fadda, with various arms penetrating the moor, so that limestone or calcareous sand load at Ballinaboy, might be transported over most of a tract about two miles long. (cited in Robinson 73)

Robinson, an English mathematician and cartographer who has learnt Irish, sees anything but 'a nameless wilderness' ripe for extraction as he climbs the 987 feet of Errisbeg. On his approach to the hill he mentions the old settlement of Cúgla, '(a name that trails more history than I can cram into a digression here)' (67). His local contact Tommy O'Donnell individualises the knolls of rock that give the hill its distinctive gap-toothed skyline, 'An Goibín Géar (meaning the small, sharp beak), Clochaí Whitewash (a slight garbling of Cloch an Whitewash, the stone of the whitewash), and Aill na gCuinneog (the cliff of the churns)' (69-70)'. Resting on a rock sheet for a moment Robinson looks out over the islands below, 'the mysterious low domes of St Macdara's Island and Deer Island further out, all these places being replete with memories and stories for me, and I planned, if time allows, I would someday write them into another book, place by place and page by page' (69).

What the Irish language has done for Robinson is to populate his landscape. The rocks, the rivers, the trees, the streams acquire a name and a personality. The human history of settlement comes alive in story and song in that language. Like Brindled Moor on the Isle of Lewis, the 'wilderness' begins to teem with memory, association and detail. If Timothy Morton believes that 'humans are traumatized by having severed their connections with nonhuman beings' (Morton 2018: 76), the Irish language, 'place by place', fills 'page by page' of pleasurable reconnection for the English learner of the Irish language. These forms of reconnection are not an idle pastime for aesthetes with time on their hands but are a vital survival skill for reimagining a different kind of environmental future. We are only likely to care about what we know and if we don't know, we won't care until it is too late to know.

One of the principal consequences of anthropogenic climate change is that humans have now become capable of affecting all life on the planet. As Dipesh Chakrabarty pointed out a number of years ago, when the collective actions of humans fundamentally alter the conditions of life on the planet, humans change from being biological agents to becoming a geological force in their own right, 'For it is no longer a question of man having an interactive relationship with nature. This humans have always had, or at least that is how man has been imagined in a large part of what is generally called the Western tradition. Now it is being claimed that humans are a force of nature in a geological sense.' (Chakrabarthy 2009: 207). With this shift in status comes a dual shift in perspective.

Firstly, it is no longer possible to speak about the 'environment' as something out there, as a negligible and dispensable externality. The environment is not exterior to, but constitutive of, who we are. Secondly, it is no longer tenable to conceive of humans as a species apart but as one species among many in relationships of increasingly acute interdependency. Therefore, we must think again about what it is to be human and if we think again about what it is to be human then we must inevitably think again about one of the activities that humans engage in, namely, language use.

Joseph Stalin, in a statement of classical historical orthodoxy in his *Dialectical and Historical Materialism* (1938), claimed that 'changes in geographical environment of any importance require millions of years, whereas a few hundred or a couple of thousand years are enough for even very important changes in the system of human society' (Stalin 1938). Stalin's distinction between natural history and human history had a certain credibility as long as the human remained in Fernand Braudel's words, a 'prisoner of climate' rather than a maker of it (cited in Crosby 1995: 1185). In the era of the Anthropocene, however, the distinction no longer holds. Once humans move from being biological agents to geological agents, dominating and determining the survival of many other species on the planet, they then become not so much subject to nature as a condition of nature itself. This dominance comes, of course, at the cost of the very survival of humanity. For this reason, trying to conceive of a sustainable future for humans means the convergence of human history with the history of life on the planet to produce a form of 'deep history'.

One crucial outcome of the emergence of the Anthropocene is a revisiting of traditional disciplinary divisions or alliances. The findings of biology and geology in the 19th century and the greatly expanded timescale of terrestrial existence that ensued led to a division of academic tasks. The geologists would concentrate on the physical history of the planet, the biologists would look at the history of organic life, and historians would devote themselves to the study of what a subsection of these organisms, humans, got up to in their time on the planet (Rudwick 2005). With the sole exception of geography, the social and human sciences increasingly defined themselves in isolation from the natural world, whether it was social and cultural anthropology under Durkheim differentiating itself from physical anthropology or psychoanalysis under Freud, considering any sensation of deep relation to the natural environment as a belated fusional fantasy from early childhood (Bonneuil and Fressoz 2013: 49-50). As Christophe Bonneuil and Jean-Baptiste Fressoz note:

> D'un côté, le territoire des sciences dures qui se firent par trop *inhumaines* avec leur conception de l'objectivité et leurs certitudes modernes; de l'autre celui des sciences de l'homme et de la société qui se firent *anti-naturelles* en mettant en avant comme le propre de l'Homme et des sociétés humaines le fait de s'arracher aux déterminations et fatalités naturelles et en conférant à "la société" sa totalité autosuffisante. (52, their emphasis)[1]

1 On the one hand, the hard sciences which became too *inhuman* with their conception of objectivity and their modern certainties; on the other, human and social sciences which became *anti-natural* by foregrounding as a distinctive feature of humans and human societies the fact of emancipation from natural fate or determination and making society a self-sufficient totality.

Tim Robinson climbing a modest hill in the west of Ireland is sensitive to these disciplinary shifts and to the radical rewriting of history in the age of the Anthropocene. Looking out from Errisbeg over the rocks in the bogs below, he claims that, 'to understand something of the nature of these rocks and of the forces that shuffled mountain ranges one across another, one has to look into births and deaths of lands and seas, the history of geography, in which, as it happens, the terrains constituting what is now Ireland have repeatedly been in the thick of the collision of continents' (Robinson 2006: 75-76). The timeline for understanding the humblest rock in a bog is immense.

In the age of human-induced climate change we need that understanding because we are no longer biological but geological agents. We need to look to the long term to understand why we have done so much damage in the short term. The Irish language, no more than any other existing human language on the planet, only goes back so far and is dwarfed by the immensity of geological time. On the other hand, it goes back far enough to create a sense of the long view, a perspective on the present informed by almost two millennia of recorded experience. One of the effects of this move towards deep history is that the familiar becomes strange. The familiar, nameless rock is given a name and a story by its name in Irish and this longer view leads us in turn to the deeper time of geology.

The landscape, taken for granted in its undescribed banality, suddenly comes alive with new information. As Timothy Morton has pointed out the word for something that is both familiar and strange at the same time is uncanny. The world around us is both familiar and, as the playing out of the climate drama indicates, deeply strange. In other words, a growing ecological awareness is bound up with this sensitivity to the uncanny. Morton argues:

> Ethics and politics might not be about tolerating, appreciating or accepting otherness. Ethics and politics might be about tolerating, appreciating or accepting *strangeness*, which boils down to *ambiguity*: how things can appear to be oscillating between familiar and strange, for example. (Morton 2018: 177 (his emphasis))

It is more than possible that the experience of the uncanny is radically underestimated for non-native speakers of Irish. For many, who have been born and grown up in Anglophone Ireland, the language can have a certain familiarity about it but, at the same time, remain disconcertingly strange. Disconcerting because of the assumption that it should come naturally, not require hours of ceaseless study and practice.

This sensation of being unsettled, of being in

an uncertain, ambiguous space (the endless soul-searching about Irish identity) can, of course, lead to resentful guilt or outright rejection. An ecological ethics based on appreciating, tolerating and accepting strangeness is, in many respects, a more realistic and more useful approach to Irish-language learning for the majority population in Ireland than the older nationalist trope of sudden, effortless conversion. What is more, this necessary humility faced with the unfamiliar becomes an ecological virtue as we unlearn many of the prejudices we had about the environment and come to a new understanding of it through the uncanny, long-term perspective of the language. Irish forces us to look at our surroundings anew and it is precisely the need to be jolted out of our complacency which becomes a pressing necessity in the midst of a climate crisis.

Attention

Relating to others, whether human or non-human, implies, first and foremost, paying attention to them. Are you all sitting still and paying attention? The familiar injunction of the school teacher has become the watchword of the new economy. If the notion of economy is typically based on the management of scarce resources, attention in a media-saturated world has become the most precious resource of all. Already by the mid-1990s Michael Goldhaber was arguing that with the emergence of digital technologies, traditional factors of production would decline in importance relative to that of attention (Goldhaber 1996; 1997). Thomas Davenport and John Beck in the *The Attention Economy: Understanding the New Economy of Business* (2001) predicted the monetisation of attention where the attention of consumers would be so sought after that they would be supplied with services free of charge in exchange for a few moments of their attention (213). We would be paid to pay attention. Google is the result. Users can now use extremely powerful search engines and all (apparently) free of charge.

From the point of view of an economics of attention, two challenges immediately present themselves. The first is how to protect attention from information overload to ensure an optimal allocation of this scarce resource (the vogue for time-management courses) and the second is how to extract the maximum amount of profit from the capture of this scarce resource (Kessous, Mellet and Zouinar 2010: 366). It is in the second sense, of course, that search engines come at a price. For Google, the user is the product, and her attention span has a lucrative exchange value. The more she pays attention, the more Google gets paid for her to pay attention. What these developments highlight is a fundamental shift in economic emphasis from production to promotion.

In information-rich environments, a series of media gates exist to filter information to potential users or consumers. Not all of these media gates have the same power co-efficient. An ad in a local college newspaper will not reach the same audience as an ad on prime-time television. If the absolute cost of diffusing information has fallen dramatically over the centuries – it is substantially cheaper to post a blog in the 21st century than to print a book in the 16th – the cost of getting past the filters of preselection has risen exponentially (Falkinger 2007: 267). In other words, as societies are more and more heavily invested in various forms of mediation, from the rise of the audiovisual industries to the emergence of digital technologies, it is less the production of goods and services than the production of demand through the capture of attention that absorbs increasing amounts of resources. Getting people to take notice is the main income generator for what McKenzie Wark has famously dubbed the 'vectorialist class', the GAFA (Google, Amazon, Facebook, Apple) oligarchy (McKenzie Wark: 2004).

There is a sense, of course, in which gaining people's attention may be a central feature of the new economy but is not necessarily novel in human experience. People have been trying to get others to sit up and take notice for millennia. As Richard Lanham points out in *The Economics of Attention: Style and Substance in the Age of Information* (2006), the central thrust of the art and science of rhetoric for more than two millennia has been to find ways of soliciting the attention of audiences. Lanham argues that much of what has been debated under the heading of 'style' in literary criticism, art history, aesthetics has largely been a matter of how writers and artists have sought to corner the attention of their readers or viewers in a field of competing media or stimuli.

Focusing on the economics of attention inevitably implies a certain set of assumptions, notably the maximisation of profits through the minimisation of costs in the context (real or imagined) of market competition. In the standard neo-classical paradigm, the economy is primarily concerned with the optimal management of scarce resources. The ends to which these resources are employed are normally outside its area of competence. However, a notion of attention which is solely concerned with means and not ends is scarcely viable as a theory of attention because attention is invariably bound up with value.

William James in his *Principles of Psychology* (1890) pointed out how a notion of attention that was purely passive was unable to account for the ways humans pay attention. James is critical of the British school of Empiricism (Locke, Hume, Hartley, the Mills, and Spencer) for not treating of the notion of 'selective attention'. He argues that because their main concern is showing that 'the higher faculties of the mind are pure products of "experience"', experience itself must be thought of as 'something simply *given*' (his emphasis). James goes on to claim:

> The moment one thinks of the matter, one sees how false a notion of experience is which would make it tantamount to the mere presence to the senses of an outward order. Millions of items of the outward order are present to my senses which never properly enter into my experience. Why? Because they have no interest for me. My experience is what I agree to attend to. Only those items which I notice shape my mind – without selective interest, experience is an utter chaos. Interest alone

> gives accent and emphasis, light and shade, background and foreground – intelligible perspective, in a word. It varies in every creature, but without it the consciousness of every creature would be a gray chaotic indiscriminateness, impossible for us even to conceive. (James 1890: 402-403)

Out of the '[m]illions of items of the outward order' we choose to pay attention to certain items and not to others. Attention inescapably involves value as attention itself implies a choice determined by particular ends (safety, sanity, satisfaction) that are believed to be important. In the circular relationship of attention and value, subjects value that to which they pay attention and pay attention to that which they value. Ends cannot, therefore, be discounted in any credible attentionscape. The purely economistic representation of attention prevents us from asking the most basic question, to what ends are directed the attention that will decide our future, or put another way, if our future is strongly determined by those things to which we might pay attention to in the present (for example, public transportation in our cities), then must not the underlying value systems of our 'selective attention' be a matter of explicit and sustained public debate?

One of the immediate effects of engaging with

another language is that you begin to pay attention to things that you previously ignored – news reports on the country where the language is spoken, songs on YouTube in that language, speakers of the language that you overhear on the Luas or the Dart or the bus. In 2013 Ethan Zuckerman, the Director of MIT's Media Lab, announced to the world an unsettling paradox (Zuckerman 2013). The Internet Age which had promised a boundless utopia of global connectivity was delivering not openness but closure. His extensive research on web and social media usage showed that users overwhelmingly accessed content in their own language, about their own culture and in their own geographical area. The electronic frontier was fast turning into the digital backyard. The political consequences of this cyber narcissism soon became all too evident in the silo hatreds of the alt-right social media, rejecting the foreign, the migrant, the impure. It meant not paying attention to other languages. Or rather it meant only paying attention to what reinforced monocultural and monolingual supremacism.

Learning another language fundamentally alters an individual's regime of attention. One could equally argue that this utopian promise extends to society as a whole. The perspectives, the insights, the understandings available on aspects of Irish environment and society in Irish available in works such as Micheál Ó Ciosáin's *Cnoc an Fhómhair* (1988), Aindrias Ó Baoill's *Cnuasach na Finne* (1993) or Seán Ó Conghaile's *Cois Fharraige le mo Linnse* (2014) are simply ignored. How social rights might relate to language rights as discussed in Aindrias Ó Cathasaigh's *An Diabhal in Uachtar in Áth Cliath* (2013) is never debated. This is why learning Irish is not some cumbersome heirloom from a remote past but is at the heart of contemporary debates on political democracy. A healthy pluralism in the public sphere is vitally dependent on expanding the range of policy and societal options on offer to deal with the climate emergency. And for this, more than ever, we need Irish.

Islands

Gilles Deleuze in the title text of a posthumous collection of essays *L'Île déserte* (2002) makes a distinction between two kinds of islands, continental islands and oceanic islands. Continental islands are islands that break off from a continental landmass, the result of a rupture or dislocation which separates them from a larger landmass – the signature note here is separation. Oceanic islands on the other hand are *sui generis*, the result of coral formations or an undersea volcano eruption. These are the originary, essential islands, islands placed under the sign of creation or recreation. What these two island types posit is two ways of imagining islands. On the one hand, the notion of island as separate, elsewhere, disjunctive, broken off. On the other, the island as the place of radical origin, where you begin all over again, where new worlds are imagined and ushered into being. In the case of both, however, Deleuze argues there is an ontological anxiety that only disappears when it is forgotten or denied. What the island foregrounds in its being is the incessant struggle between land and sea, islands defined by the sea just as the land beneath the sea may emerge at any moment. There is a fragile equilibrium between island and ocean so that in Deleuze's words, 'L'homme ne peut bien vivre, et en sécurité, qu'en supposant fini (du moins dominé) le combat vivant de la terre et de l'eau' (Deleuze 2002: 11) [Humans can only live well and securely assuming that the struggle between land and sea is over (or at the very least contained)].

There are two ways to consider the Deleuzian analysis in the light of recent events in these islands. Firstly, we can see how the notion of the continental island can feed into a Brexit imaginary which oscillates between the poles of separation and recreation. Breaking away from the European Union is a paradoxical recognition of the continental nature of the island of Britain, but continental in its separateness not in its conjoinedness. Secondly, post-Brexit Britain becomes for Leave voters, both on the Right and the Left, the opportunity for beginning all over again, restoring the pristine possibility of utopia to the island. Leaving it all behind means starting all over again. For Ireland, on the other hand, matters are more complex, demanding a partial redefinition of Deleuzian insularity. When sea and land levels stabilised after the last ice age and rising sea levels flooded coastal areas, severing the land bridges connecting us to Britain and Europe around 7000 years ago, the island of Ireland emerges. Ireland is a continental island but an island that seeks not separation from but unity with the European continent. And it is arguably for this reason that the second insular function will have to come into play, namely that of the imaginative exploration of future possibility. It is in this respect that Deleuze adds an important nuance to his island discussion when he claims that the appearance of the island is *the* original creation and what occurs on the island is not the 'origine' but the 'origine seconde' (16), the 'naissance' of the island gives birth to the 'renaissance' of the culture. As Deleuze notes, we characterise life by its capacity to reproduce itself. A once-off production would be an anomaly. It is only repetition, the series, that indicates we are in the presence of the living. Taking that to be the case then, we can argue that what is of crucial importance for Ireland heading into the post-Brexit world (given that the majority of its inhabitants north and south voted to remain in Europe) is that it reimagines the island not as the dead, originary site of ethno-national purity but as a living site of 're-création', 're-commencement'. This is where, I believe, to paraphrase the title of a book by the former Archbishop of Canterbury, Rowan Williams, there is room for the Irish Language in the Public Square. In all the current debates around Europe and Brexit where we have historians, economists, political scientists furiously debating past and future scenarios for Ireland in Europe, there is a remarkable absence of contributions from those in the public sphere, despite the fact that one might suppose that if anyone knew about what it might be like to be attached or related to the European continent it would be writers and scholars who have engaged and written about this connection for centuries.

As we mentioned earlier, in our discussion of Deleuze's views on islands, there is the founding tension of the island, centred around 'le combat vivant de la terre et de l'eau'. For Jean-Claude Guillebaud in his *Un voyage en Océanie* (1980), this tension explains islanders' intense attachment to place – it might disappear at any moment – and their incurable nomadism – the waters all round enclose them in their island prison, so best to get out while the going is good. Here I want to look more closely at insularity and sovereignty and ask whether the French theorist is not reminding us of a tragic oversight that will cost us dear in terms of the 'retour de l'humide' that will fatally compromise our futures. In the ceremonies surrounding the centenary of the 1916 Rising three years ago, a phrase often repeated was the sentence beginning 'We declare the right of the people of Ireland to the ownership of Ireland, and to the unfettered control of Irish destinies, to be sovereign and indefensible'. This is a classic nationalist statement of territorial sovereignty but as Deleuze might say does it hold water, or more pointedly, what does it do with water?

Ireland has a coastline of approximately 7400km. As the Centre for Climate Change Adaptation puts it, 'Most of the Irish Sea coast is experiencing long-term recession, beaches and dunes are retreating.' (Marei

2019). Each year two kilometres of land in 300 sites around the coast are lost due to erosion and flooding, and this rate is set to increase dramatically with global sea-level rise. Kevin O'Sullivan commenting on a report by the UN Intergovernmental Panel on Climate Change noted that the negative impacts of greenhouse-gas emissions on the world's oceans, 'are likely to happen whether climate-heating emissions are curbed or not in coming decades, while many millions more of people living in coastal communities – including Ireland – face more extreme flooding and sea-level rise' (O'Sullivan 2019: 2). In other words, although much attention has been focused on territorial issues around Brexit, there is a startling lack of attention to the ecological dimension to sovereignty. There is no point in claiming the right to ownership of land that is no longer there. Which in turns suggests that we need to perhaps revisit the link between territory and sovereignty, and in the words of Bruno Latour in *Face à Gaïa: huit conférences sur le nouveau régime climatique*, think about the need for *souveraineté partagée* or joint sovereignty (Latour 2015: 358). This is power-sharing in the sense where we look at non-human agents (water/temperature/soil) as constituent parts of our common world. This is a kind of post-human, post-nationalist sovereignty that sees territorial integrity based not on separation and exclusion but on interdependence and inclusion.

However, in order to develop this notion of ecological sovereignty which takes territorial integrity seriously in an environmental sense, we must reintegrate a sense of island identity. By identity here, we do not mean the sense of a shared ethnic commonality, a notion which is deeply problematic at any moment, but rather a shared concern for the geophysical integrity of the Atlantic island that is our home. It is time to come back down to earth. Not easy for a population that spent almost €5 billion on foreign holidays in 2018 and recorded 8.6 million outbound trips, an increase of 5.8% on 2017 (Pope 2019). Taking flight from our insular reality stores up inevitable trouble for the future. Any notion of sustainable travel or transport has first and foremost to address the questions of political geography if it is to have any hope of translation into policy. By this we mean that it is only where there is a renewed engagement of the public imagination with the reality of Ireland as island will there be a democratic space receptive to the necessary changes of ecological adaptation.

When we look to where we might find the necessary imaginative resources to trigger this engagement, one obvious candidate is the body of language and story which has been profoundly shaped for more than two thousand years by the island experience. When Darach Mac Con Iomaire directed a multi-media show to celebrate *Bliain na Gaeilge* in 2018 the central motif linking the different periods and fortunes of the language was a *currach* in movement. The texts and songs threading through the performance spoke of the constant tension, the endless renegotiated relationship between land and sea on the island of Ireland. The excerpt from Máirtín Ó Cadhain's *Cré na Cille* which graced the back cover of the programme notes was a terse summary of this island story, 'Tá an fáiteallaí ag cur chun farraige le port ar a bhéal a bhfuil spreacadh taoile, gaoithe agus gréine ann' [The forager heads out to sea with a song on his lips, energised by sea, wind and sun]. This is not the manifesto of a soft-focus romanticism, wending the wild Atlantic way of photogenic cliché, but an invitation to engage with two millennia of thinking about insular realities that are a key component of our ecological survival. From Amergin Glúingel to Nuala Ní Dhomhnaill, writing in Irish has constantly probed the many dimensions – human and natural – to our island identity. It is this imaginative engagement with Ireland as island, in a spirit of ecological humility rather than annexationist triumphalism, which has the potential to make the Irish language matter in new and unexpected ways. The critic and writer Robert Welch once famously argued that:

> In questions of culture and tradition everything comes back to language.

> Whenever there is a crisis, of something vital being transacted, the words a person uses, in speech or in writing, become crucial. (Welch 1993: 32)

We are now in that moment of crisis. Something vital for our future is being transacted. More than ever the words, the languages, we choose to use and value are crucial. Choosing Irish as one of those languages is essential in averting that crisis. Otherwise, there is no future, only monocultures of extinction.

References

Abley, Mark (2001) *Spoken Here: travels among threatened languages*, London: Heinemann.

Arthur, Chris (2019) 'Escaping Indoors', *World Literature Today*, summer, 60-64.

Beck, John and Davenport, Thomas (2001) *The Attention Economy: Understanding the New Currency of Business*, Cambridge (Mass.): Harvard Business School.

Bonneuil, Christophe and Fressoz, Jean-Baptiste (2013) *L'Événement Anthropocène: La Terre, l'histoire et nous*, Paris: Seuil.

Berry, Wendell (2000) *Life as a Miracle*, Berkeley (CA): Counterpoint Press.

Brüscher, Bertram (2013) 'Nature on the Move: The Value and Circulation of Liquid Nature and the Emergence of Fictitious Conservation', *New Proposals: Journal of Marxism and Interdisciplinary Inquiry*, 6, 20-36.

Campbell, Anne (2013) *Rathad an Isein: The Bird's Road – a Lewis Moorland Glossary*, Glasgow: Farram.

Carruthers, Peter (2009) 'The Land Debate – "Doing the Right Thing": Ethical Approaches to Land-Use Decision-Making', 294-318, in Winter, Michael and Lobley, Matt (eds.) *What is Land For? The Food, Fuel and Climate Change Debate*, London: Routledge.

Casey, Donal (2018) 'Is the Irish language dead or alive?' *IrishCentral*, 26 March. Available online: https://www.irishcentral.com/roots/history/irish-langauge-alive-dead (accessed 3 September 2019).

Central Statistics Office, *Environmental Indicators Ireland 2019*. Available online: https://cso.ie/en/releasesandpublications/ep/p-eii/eii19/mainfindings/ (accessed 28 August 2019)

Conservatives (2015) *The Conservative Party Manifesto 2015*. Available online: https://s3-eu-west-1.amazonaws.com/manifesto2015/ConservativeManifesto2015.pdf (accessed 6 September 2019).

Crosby, A.W. (1995) 'The Past and Present of Environmental History', *American Historical Review*, 100, October, 1185.

Danowski, D. and Viveiros de Castro, E. (2014) 'L'Arrêt de monde', 221-339, in Émilie Hache, *De l'univers clos au monde infini*, Bellevaux : Éditions Dehors.

Daunt, William Joseph O'Neill (1848) *Personal Recollections of the Late Daniel O'Connell M.P.*, London: Chapman and Hall,

Delisle, Jean and Woodsworth, Judith (2012) *Translators through History*, revised edition, Amsterdam: John Benjamins.

Deleuze, Gilles (2002) *L'Île déserte*, Paris: Éditions de Minuit.

Department of Communications, Climate Change and the Environment (2019) *Climate Action Plan 2019*. Available online: https://www.dccae.gov.ie/documents/Climate%20Action%20Plan%202019.pdf (accessed 31 August 2019).

Falkinger, John (2007) 'Attention Economies', *Journal of Economic Theory*, vol.133, 266-294.

Friel, Brian (1981) *Translations*, London: Faber.

Fogarty, Pádraig (2017) *Whittled Away: Ireland's Vanishing Nature*, Cork: Collins.

Garvin, Tom (2005) *Preventing the Future: Why was Ireland so poor for so long?* Dublin: Gill.

Goldhaber, Michael (1996) 'Principles of the New Economy'. Available online; well.com (accessed 27 September 2019).

Goldhaber, Michael (1996) 'Some Attention Apothegms'. Available online: well.com (accessed 14 August 2019)

Goldhaber, Michael (1997) 'Attention Shoppers!', *Wired*, 12, 5. Available wired.com (accessed 15 August 2019)

Goldhaber, Michael (1997) 'The Attention Economy and the Net', *First Monday*, vol.7, no.2. Available online: FirstMonday.org (accessed 16 August 2019).

Guillebaud, Jean-Claude (1980) *Un voyage en Océanie*, Paris: Seuil.

Heller, M. and Duchêne, A. (2007) 'Discourses of endangerment: Sociolinguistics, globalization and social order', in Duchêne, A. and Heller, M., eds., *Discourses of endangerment: Ideology and interest in the defence of languages*, London: Continuum, 1-13.

Hyde, Douglas (2019) [1892] 'The Necessity for De-Anglicising Ireland'. Available online: https://www.thefuture.ie/wp-content/uploads/1892/11/1892-11-25-The-Necessity-for-De-Anglicising-Ireland.pdf (accessed 1 September 2019).

Intergovernmental Science-Policy Platform on Biodiversity and Ecosystem Services (2019) *Global Assessment Report*. Available online: https://www.un.org/sustainabledevelopment/blog/2019/05/nature-decline-unprecedented-report/ (accessed 4 September 2019)

James, William (1890) *The Principles of Psychology*, New York: Henry Holt.

Kelman, Ari (2019) 'The myth of the Vanishing Indian', *The Times Literary Supplement*, July 5, no. 6066, 12.

Routledge.

Kessous, E., Mellet, K. and Zouinar, M. (2010) 'L'économie de l'attention. Entre protection des ressources cognitives et extraction de la valeur', *Sociologie du travail*, vol.52, no.3, 359-373.

Lanham, Richard (2006) *The Economics of Attention: Style and Substance in the Age of Information*, Chicago: University of Chicago Press.

Klein, Naomi (2014) *This Changes Everything: Capitalism vs. the Climate*, London: Allen Lane

Morton, Timothy (2018) *Being Ecological*, London: Pelican.

Latour, Bruno (2015) 'Telling Friends from Foes in the Age of the Anthropocene'. Available online: https://www.google.ie/?gws_rd=ssl (accessed 14 September 2019).

Latour, Bruno (2015) *Face à Gaïa: huit conférences sur le nouveau régime climatique*, Paris: La Découverte.

Latour, Bruno (2017) *Où atterrir? Comment s'orienter en politique*, Paris: La Découverte.

Marei (2019) 'Climate Change and Adaptation'. Available online: https://www.marei.ie/research/climate-change-and-adaptation (accessed 2 September 2019).

McIntosh, Alastair (2008) *Hell and High Water: Climate Change, Hope and the Human Condition*, Edinburgh: Birlinn.

Morley, Vincent (2017) *The Popular Mind in Eighteenth Century Ireland*, Cork: Cork University Press.

Ní Chinnéide, Mairéad (2013) *Scéal Ghael-Linn*, An Spidéal: Cló Iar-Chonnacht.

Ó hÓgáin, Dáithí (1990) *Myth, Legend, Romance: An Encyclopaedia of the Irish Folk Tradition*, London: Ryan.

O'Sullivan, Kevin (2019) 'Climate change severely damaging world's oceans, UN report warns', *The Irish Times*, 26 September, 2.

Palmer, Patricia, Baker, David J. and Maley, Willy (2019) 'Enter MacMorris', *Dublin Review of Books*, 113, July. Available online: https://www.drb.ie/essays/enter-macmorris (accessed 29 August 2019).

Plumwood, Val (1994) 'Nature, Self and Gender: Feminism, Environmental Philosophy, and the Critique of Rationalism', 142-59, in Gruen, l. and Jamieson, D. (eds) *Reflections on Nature: Readings in Environmental Philosophy*, Oxford: Oxford University Press.

Pope, Conor (2019) 'Irish residents spent almost €5 bn on foreign holidays in 2018 – CSO', *The Irish Times*, June 19. Available online: https://www.irishtimes.com/life-and-style/travel/irish-residents-spent-almost-5bn-on-foreign-holidays-in-2018-cso-1.3931117 (accessed 29 September 2019).

Rahbi, Pierre (2010) *Vers la sobriété heureuse*, Arles: Babel.
Robinson, Tim (2006) *Connemara: Listening to the Wind*, Dublin: Penguin Ireland.
Rymer, Russ (2012) 'Vanishing Voices', *National Geographic*, July. Available online: https://www.nationalgeographic.com/magazine/2012/07/vanishing-languages (accessed 20 September 2019).
Salmond, Anne (2017) *Tears of Rangi: Experiments across Worlds*, Auckland: University of Auckland Press.
Spenser, Edmund (1970 [1596]) *A View of the Present State of Ireland*, ed. W.L. Renwick, Oxford: Clarendon.
Stalin, Joseph (1938) *Dialectical and Historical Materialism*. Available online: https://www.marxists.org/reference/archive/stalin/works/1938/09.htm (accessed 3 August 2019).
Stibbe, Arran (2015) *Ecolinguistics: Language, ecology and the stories we live by*, London and New York.
Treuer, David (2019) *The Heartbeat of Wounded Knee: Native America from 1890 to the present*, London: Corsair.
Virilio, Paul (2009) *Le futurisme de l'instant: stop-eject*, Paris: Galilée.
Welch, Robert (1993) *Changing States: Transformations in Modern Irish Writing*, London: Routledge.
Wilson, Edward (2003) *The Future of Life*, London: Abacus.
Wolf, Nicholas M. (2014) *An Irish Speaking Island: State, Religion and the Linguistic Landscape in Ireland 1770-1870*, Madison: University of Wisconsin Press.

Wong, R. (2013) 'The Oxygen Trade: leaving Hondurans gasping for air', *Foreign Policy in Focus*, 18 June.

Yildiz, Yasemin (2012) *Beyond the Mother Tongue: The postmonolingual condition*, New York: Fordham University Press.

Zuckerman, Ethan (2013) *Digital Cosmopolitans: Why We Think the Internet Connects Us, Why It Doesn't, and How to Rewire It*, New York: Norton.

An Ghaeilge agus an Éiceolaíocht

Michael Cronin

Foilseacháin
Ábhair
Spioradálta

Foras na Gaeilge

Foilseacháin Ábhair Spioradálta
Baile Átha Cliath

An chéad chló 2019

Foilseacháin Ábhair Spioradálta

Cé go bhfuil an téama céanna faoi chaibidil sa dá chuid den leabhar seo, ní mar a chéile léiriú ná ionramháil a bhfuil iontu.

Foras na Gaeilge

Tá Foilseacháin Ábhair Spioradálta buíoch de Fhoras na Gaeilge as tacaíocht airgid a thabhairt dóibh.

Dearadh clúdaigh: Fionnuala Mac Aodha & Anthony Bools

Clóchur: Messenger Publications
Clódóirí: W&G Baird Ltd

ISBN 978-1-906982-66-9

Clár

Oidhreacht

I Samhain 2018 d'fhoilsigh Rialtas na hÉireann plépháipéar dar teideal *Oidhreacht Éireann 2030: Comhairliúchán Poiblí.* An sprioc a bhí ag an gcáipéis ná tús a chur le díospóireacht phoiblí ar thodhchaí na hoidhreachta in Éirinn. I mbrollach an phlépháipéir déanann an tAire Cultúir, Oidhreachta agus Gaeltachta iarracht coincheap na hoidhreachta a shainmhíniú:

> Thart orainn ar an uile bhealach atá ár n-oidhreacht. Baineann sí leis an am atá thart ó thaobh ár séadchomharthaí agus lonnaíochtaí ár sinsear. Baineann sí leis an am i láthair ó thaobh na tírdhreach éagsúil álainn a gcónaímid iontu sa lá atá inniu ann. Is tírdhreacha iad sin lenar mealladh ár sinsir [le] cur fúthu anseo na mílte bliain ó shin agus a gcothaíonn a mbithéagsúlacht sinn sa lá atá inniu ann fós. Baineann sí leis an am atá romhainn sa dóigh is gur leis na cinn a dhéanaimid sa lá atá inniu ann a mhúnlófar oidhreacht ár leanaí agus na nglún atá romhainn. (An Roinn Cultúir, Oidhreachta agus Gaeltachta 2018: 3)

Nuair a mhíníonn an tAire an fhís atá aici féin agus ag an Roinn, maíonn sí go bhfuil sí simplí agus d(h)úshlánach, 'go ndéanfaimís ár n-oidhreacht a shainaithint agus a chosaint ar an mbealach is fearr is féidir linn' (3). Ach conas is féidir linn an oidhreacht seo a shainaithint muna bhfuil na huirlisí cuí againn? Conas is féidir í a chosaint muna bhfuilimid in ann cruth nó brí na hoidhreactha sin a thuiscint? Luaitear 'séadchomharthaí agus lonnaíochtaí ár sinsear' ach gan focal ann faoin teanga a bhí á labhairt ag an gcuid ba mhó dár sinsear. Luaitear an tírdhreach 'álainn éagsúil' ach níl trácht ar bith ar an teanga a thug stair, pearsantacht agus brí don tírdhreach céanna.

Níos faide siar sa cháipéis maítear go gcothaíonn 'éagsúlacht agus saibhreas inár n-oidhreacht ár mbraistint chomhuintearais, folláine, féiniúlachta agus áite' (6). Ach cá bhfuil an éagsúlacht agus an saibhreas seo le feiceáil agus gan aon chaint ar chaint, ar na focail, atá mar nasc beo idir na glúnta agus mar fhoinse féiniúlachta do phobal ar bith agus é ag iarraidh a suíomh sa timpeallacht a thuiscint? Deirtear linn go ndíreoidh *Oidhreacht Éireann* 'ar ár n-oidhreacht thógtha agus nádúrtha' (9) ach conas is féidir teanga, gléas cumarsáide agus tusiceana an duine aonair, a scaradh amach ó stair agus ó chiall na hoidhreachta agus na timpeallachta? Níl focal ann ach an oiread sa phlépháipéar ar athrú aeráide, an baol

is mó don 'bithéagsúlacht' atá mar ghné shuntasach, dar leis an Aire, de thírdhreach na tíre. Ní haon ionadh, ar bhealach, go bhfuil ról na teanga agus ceist na héiceolaíochta ar lár sa cháipéis oifigiúil seo, mar léiríonn an neamhaird sin easpa díospóireachta agus plé sa phobal i gcotinne maidir leis an cheangal idir cúrsaí teanga agus cúrsaí timpeallachta.

Fiú i gcáipéisí a bhaineann go díreach le feidhn agus todhchaí na Gaeilge fanann an Roinn ina tost maidir leis an teagmháil idir cleachtais teanga agus an ghéarchéim éiceolaíoch. Níos luaithe in 2018 tháinig cáipéis eile amach ón Roinn Cultúir, Oidhreachta agus Gaeltachta dar teideal *Plean Gníomhaíochta 2018-2022: Straitéis 20 Bliain don Ghaeilge 2010-2030*. Murab ionann is an foilseachán eile ón Roinn aithnítear ról stairiúil na teanga ann:

> Tá an Ghaeilge ar cheann de na teangacha is sine atá go fóill in úsáid san Eoraip agus stair shaibhir chultúrtha ag dul léi. Is í an Ghaeilge príomhtheanga na hÉireann agus, mar sin, is mionteanga neamhghnách í ar go leor bealaí sa mhéid is gur tugadh cosaint shuntasach bhunreachtúil agus reachtach di ó bunaíodh an Stát. (An Roinn Cultúir, Oidhreachta agus Gaeltachta 2018a: 11)

Níl aon mhíniú ann, áfach, ar fhiúntas comhaimseartha na teanga agus ar an 'stair shaibhir chultúrtha' atá ag gabháil léi. Deirtear freisin, 'is cuid riachtanach d'oidhreacht bheo an Stáit í an Ghaeilge agus is acmhainn thábhachtach í an Ghaeltacht ar chóir gach iarracht a dhéanamh í a chosaint, a chaomhnú agus a fhás' (11). Arís, níl leid dá laghad ann maidir don cheangal idir an 'oidhreacht bheo' seo agus an Ghaeilge. Sa 'Ráiteas' ó Aire eile, an tAire Stáit don Ghaeilge, don Ghaeltacht agus do na hOileáin, ní airímid aon cheo faoin ngéarchéim éiceolaíoch, ná faoi thábhacht na teanga chun dul i ngleic leis an ngéarchéim chéanna. Is cuid de Straitéis 20 Bliain don Ghaeilge an Plean Gníomhaíochta, ach cén sórt pláinéid a bheas againn i 2030? Agus Éire mar cheann de na tíortha is measa san Eoraip maidir le hastaíochtaí carbóin, níl faic sa Phlean Gníomhaíochta faoin dlúthcheangal idir teanga agus timpeallacht.

Agus na mílte páiste scoile amuigh ar na sráideanna ag agóidíocht ar son polasaithe réadúla chun déileáil leis an athrú aeráide, cén fhís atá ag an Roinn Oideachais agus Scileanna agus í ag réiteach polasaí do Straitéis 20 Bliain don Ghaeilge? I gcáipéis a chuireann síos go mion ar na cuspóirí straitéise atá ag an Roinn maidir leis an teanga níl focal ar bith in aon áit faoin timpeallacht. Mar shampla, ag

tús na cáipéise mínítear an fhís atá ag an Roinn i leith na Gaeilge agus deirtear gur aidhm pholasaí an rialtais, 'a chinntiú go mbeidh an Ghaeilge níos infheicthe sa tsochaí againn, mar theanga labhartha ag ár gcuid saoránach mar aon le bheith le feiceáil ar chomharthaí agus i litríocht' (An Roinn Oideachais agus Scileanna 2009: 2). Ach tá an Ghaeilge le feiceáil chuile lá den bhliain sna hainmneacha, agus sa stair agus sna nósanna a bhaineann le timpeallacht nádúrtha na hÉireann. Is léir go raibh an Roinn dall ar an ngné seo den teanga agus is follas ón *Tuarascáil ar Dhul Chun Cinn: 2010 go 2015* nach raibh aon athrú intinne ann. Is beag an dul chun cinn atá déanta ó thaobh na smaointeoireachta de. Fiú agus a thodhchaí ag brath ar mheon iomlán difriúil i leith na timpeallachta in Éirinn, níl an t-aos óg nó an dream atá freagrach as a gcuid oideachais ag fáil treoir ar bith ó na heagrais Stáit atá in ainm is a bheith ag obair lena leas.

Closimid anois, ámh, guthanna nua a léiríonn dearcadh níos caolchúisí agus níos ábhartha i leith na teanga. Ina bplépháipéar a foilsíodh faoi choimirce Acadamh Ríoga na hÉireann faoin teideal *An Ghaeltacht, An Ghaeilge, An Béaloideas agus Cruthaitheacht na nDaoine*, taispeánann Angela Bourke agus Diane Negra gur chóir léamh éiceolaíoch a chur chun cinn agus cás na Gaeilge à

plé ag tráchtairí polaitiúla agus cultúrtha. Feiceann siadsan an ceangal láithreach idir teanga agus timpeallacht:

> Na tírdhreacha drámatúla gan chónaí atá fairsing sa Ghaeltacht, is léiriú iad ar leochaileacht na timpeallachta: éiceachóras sléibhe agus portaigh, locha agus cladaigh, ina maireann ainmhithe agus plandaí tearca ar scáth a chéile, áit a dtéann damáiste d'eilimint amháin i bhfeidhm ar an iomlán. Tá éiceachóras na Gaeilge chomh sainiúil agus chomh leochaileach céanna. (Bourke agus Negra 2019: 7)

Déantar an chomparáid seo, áfach, ní chun cur síos a dhénamh uair amháin eile ar an gcruatan gaelach ach chun saibhreas na Gaeltachta agus na Gaeilge a léiriú i bhfriotal na héiceolaíochta. Dar leis na húdair, is gá an Ghaeltacht a aithint, mar shampla, mar éiceachóras sóisialta agus cultúir, ní chun í a chosaint mar eisceacht ach chun í a chothú mar eiseamláir éiceolaíoch don chuid eile den tír. Mar a mhaíonn Bourke agus Negra, freagraíonn fiontair Ghaeltachta don smaointeoireacht nua atá ar fáil i réimse an ghnó maidir le luachanna atá bunaithe ar fhís níos leithne

ná cúrsaí airgid amháin, 'radharc fadtéarmach, muintearas buan, fréamhacha áitiúla, teorainneacha aitheanta, pobal foghlama agus freagracht ceannairí' (7). Ag trácht ar an mbéaloideas agus ar na síscéalta, deir siad:

> Tá córas ollmhór samhlaíochta taobh thiar de na síscéalta seo, atá ceangailte le chéile i ngréasán taisce, ina bhfuil eolas faoi scileanna, faoin tíreolaíocht, an aimsir, an sláinteachas (fisiciúil, sóisialta agus intinne), agus go leor d'imeachtaí an chine dhaonna[...]I saol na scéalaíochta, is timpeallachtaí oibre iad an tírdhreach agus an fharraige, ach is léarscáileanna comhroinnte iad freisin de dhomhan na cuimhne, na mothúchán agus na samhlaíochta. (11)

Mar sin, is follas 'go bhfuil níos mó i gceist le teanga ná seasamh polaitiúil, agus go bhfuil saibhreas tuisceana agus samhlaíochta le tairiscint leis an nGaeilge' (18). B'fhéidir go bhfuil seasamh polaitiúil i gceist anseo agus an teanga á nascadh le marthanas na timpeallachta ach ní hé an seandíoscúrsa náisiúnaíoch atá ann ach polaitíocht nua dhomhanda a cheanglaíonn meas agus tusicint

ar an dúchas le marthanas an phláinéid.

Sa cháipéis ón Roinn Oideachais a luamar cheana tá focal tábhachtach ann ag deireadh abairt faoi aidhmeanna an rialtais, is é sin an focal 'litríocht'. Agus muid ag moladh dhomhan 'na cuimhne, na mothúchán agus na samhlaíochta' ní mór dúinn machnamh a dhéanamh freisin ar shuíomh na litríochta in éiceachóras na Gaeilge. I dtuarascáil faoi *Dearcadh an Phobail i leith na Gaeilge ar oileán na hÉireann* deir na húdair san Achoimre Fheidhmeach gur 'I bPoblacht na hÉireann, is gnách go measann daltaí bunscoile agus iar-bhunscoile nach bhfuil an Ghaeilge chomh spéisiúil ná chomh húsáideach le hábhair eile, an Béarla nó an Mhatamaitic, cuir i gcás' (Darmody agus Daly 2015: ix). Seanphort é seo a théann siar go haimsir an Drochshaoil agus an deacracht maidir le muc a dhíol as Gaeilge. Cad is fiú an teanga sin a fhoglaim? Tá an teanga in áit na leathphingine toisc nach bhfuil pinginí le fáil aisti. Ach de réir an staidéir ní dóigh le páistí scoile áirithe go bhfuil an Ghaeilge spéisiúil ach an oiread. An réiteach ar an scéal seo, dar le tráchtairí áirithe, ná go bhfuil ualach na litríochta ar an gcuraclam Gaeilge iomarcach agus go gcaithfear fáil réidh leis an mbéim seo ar an bhfocal scríofa agus cloí leis an teanga labhartha díreach mar a rinneadh leis an gcuraclam Fraincise i meánscoileanna Phoblacht na hÉireann.

Ar bhealach, ní haon ionadh go bhfuil amhras faoi luach na Gaeilge agus drochmheas ar an litríocht fite fuaite le chéile in intinn an-chuid scoláirí, mar baineann an ceangal coitianta seo le fréamhacha na géarchéime éiceolaí féin. An meon atá taobh thiar de loscadh an phláinéid ná an meon fóntaíoch. An t-aon fheidhm atá le garraí nó coill nó ainmhí nó leis an bhfarraige ná mar fhoinse brabúis. Sin é go díreach dúshlán na litríochta Gaeilge. Bréagnaíonn sí an fhealsúnacht éadomhain atá taobh thiar den dearcadh santach seo mar go dtaispeánann sí arís agus arís eile tábhacht na háite ina bhfuilimid lonnaithe. An tuairim a bhí ag Seán Ó Tuama, mar shampla, nach raibh traidisiún liteartha eile in iarthar na hEorpa a raibh oiread tábhachta leis an áit féin ann agus atá i dtraidisiún na hÉireann (Ó Tuama 1992: 176-185). Scoláire eile a bhí ar aon intinn leis ná Brian Ó Cuív nuair a mhaígh sé go raibh logainmneacha, mar shampla, lárnach i bhforbairt na litríochta Gaeilge in Éirinn ar feadh na gcéadta bliain (Ó Cuív 1989-90: 90-106). Ní haon ionadh go raibh bá ag lucht an phinn le seanchas áiteacha mar tugann teanga treoir do dhaoine maidir le leithleachas na háite. Deir Roibeard Ó Maolalaigh an méid seo a leanas:

> Tá fios fada is fairsing gur de dhlúth
> is d'inneach chultúr is chosmeolaíocht

> na nGael coincheap na "háite", na logainmneacha a thugtar ar áiteacha is an dinnsheancas a thógtar mórthimpeall orthu. Do na Gaeil úd go háirithe a bhaineann a mbeatha amach ag obair ó lá go lá ar an talamh agus ar an bhfarraige, bíonn tábhacht nach beag ag baint le logainmeacha atá intuigthe don phobal a úsáideann iad is a chuireann treoir shiombalach, chruinn ar fáil dóibh agus iad i mbun a mbeatha laethúla. Mar a dúirt Seán Choilm Chúláin as Carna: "Mara bhfuil aon ainmeachaí ar na garrantaí níl tú in aon áit" (Ó Maolalaigh 2003: 195)

Ach tá an scil sin ag teastáil ó chuile dhuine i láthair na huaire agus ní hamháin ón dream atá ag obair 'ar an talamh agus ar an bhfarraige'. Ní foláir teacht 'ar threoir shiombalach, chruinn' arís mar go bhfuil inbhuanaitheacht na timpeallachta ag brath go láidir ar an tuiscint agus an meas atá ag daoine ar an áit ina bhfuil siad ina gcónaí. Ar ndóigh, is féidir teacht ar eolas beacht fisicúil faoi áit tríd an ainm. Luann P.W. Joyce Tráigh Bhréagach (*Trawbreaga*) i Málainn i gCo. Dhún na nGall mar a bhfuil trá chontúirteach, 'for the tide rises there so suddenly that it has often swept away people walking incautiously on the

shore' (Joyce 1883: 437). Baineann ríthabhacht leis an saibhreas atá folaithe i logainmneacha agus i ndinnseanchas, áfach, mar cruthaíonn sé ceangal idir samhlaíocht agus stair an duine agus an domhan atá mórthimpeall air. Tagann meas ar áit as an gceangal seo – "Mara bhfuil aon ainmeachaí ar na garrantaí níl tú in aon áit".

Is é croílár na faidhbe maidir leis an ngéarchéim éiceolaíoch ná go bhfuilimid tar éis dul ar strae. Nílimid in aon áit nó ba chóir dúinn a rá nach bhfuilimid in aon áit cheart. Toisc go bhfuilimid dall ar an áit mar áit faoi leith le stair, le cumhacht filíochta agus tuisceana, tugaimid neamhaird ar an dochar leanúnach atá á dhéanamh don áit ag gníomhaíochtaí áirithe daonna. Cuidíonn litríocht na Gaeilge linn chun teacht arís ar an eolas riachtanach, imeallaithe seo. Ag trácht dó ar dhán a chum Séamus Ó Doraidheáin san ochtú haois déag, 'Sliabh Liag', ag tabhairt ómóis do shliabh i dTír Chonaill, deir Lillis Ó Laoire, scoláire agus amhránaí, gur dea-shampla é de phíosa scríbhneoireachta a naisceann na traidisiúin éagsúla i saíocht na nGael – seanchas an tslua sí agus seanchas na Críostaíochta – agus 'béithe na Gréige ag seinm ar uirlisí Gaelacha' (Ó Laoire 2003: 48). Cantar an dán seo mar amhrán fós i dTír Chonaill agus, dar le Lillis Ó Laoire, léiríonn traidisiún na hamhránaíochta Gaeilge cé chomh

tábhachtach is atá an dinnseanchas mar bhuncloch na samhlaíochta. Ní hamháin sin ach caithfimid éisteacht go géar leis an teachtaireacht phráinneach atá taobh thiar de na hamhráin chéanna:

> Sa lá inniu, tá formhór phobal na hÉireann ina gcónaí i mbailte móra, in áiteacha a bhfuil eastáit á dtógáil agus á n-ainmniú le fada, agus mí-úsáid de gach cineál á baint as an talamh. Beag beann ar leanúnachas stairiúil agus cultúrtha, is dóigh liom go bhfuil tábhacht i gcónaí leis na teachtaireachtaí atá le fáil sa litríocht a cheiliúras ómós áite ar bhealach atá chomh neamhleithscéalach, spleodrach sin. Más fíor gur lámhscríbhinn é an tírdhreach nach bhfuil ar chumas lucht an Bhéarla in Éirinn í a léamh feasta[...]is dóigh go bhfuil ár gcóras pleanála agus ár bpolaiteoirí chomh neamhoilte céanna is atá lucht na litríochta. (62)

Níl aon amhras ar Ó Laoire ná go bhfuil dlúthbhaint idir loitmhilleadh timpeallachta agus an t-aineolas teanga. Cothaíonn bochtanas teanga bochtanas intinne. Ní féidir an teanga a scaradh amach ón lón samhlaíochta agus tusiceana atá le fáil inti. Ach ní

rud teibí í an teanga ná an litríocht. Bíonn torthaí nithiúla ag chuile ghníomh teanga. Muna bhfuiltear in ann an tírdhreach a léamh cad a tharlaíonn? Esapa measa ar an áit, scrios éiceolaíoch, drochphleanáil agus faillí i leith fhorbairt inbhuanaitheach an oileáin. Má bhíonn páiste scoile ar bith ag lorg fianaise chun a thaispeáint cé chomh húsáideach is atá an Ghaeilge ní gá dó nó di ach amharc mórthimpeall ar an timpeallacht atá á scrios ar bhonn laethúil in Éirinn. Mar a bhí scríofa ar fhógra agus é á iompar ag duine amháin de na páistí sin ar agóid i mBaile Átha Cliath, 'Ní thiocfaidh ár lá'. Agus cinnte ní thiocfaidh má leanaimid ar aghaidh lenár ngnáthchleachtais teanga ar an oileán.

Nádúr-Fhilíocht

Scríbhneoir a thuig cé chomh luachmhar is atá litríocht na Gaeilge chun teacht i dtír ar ghnéithe éiceolaíocha de shaol agus d'eispéireas na nÉireannach ná Flann O'Brien (Brian Ó Nualláin). *Nádúir-fhilíocht na Gaedhilge: tráchtas maraon le duanaire* an teideal a bhí ar an leagan deireanach den tráchtas MA a scríobh sé ag tosnú in 1934, agus bronnadh an chéim air an bhliain dár gcionn (de Paor 2018: 44-70). Cuireann sé béim sa tráchtas seo ar neamhspleáchas na bhfilí Gaeilge a ghlac le leithleachas an nádúir:

> Ní raibh fonn ar mheon ceilteach cumhacht Dé a fheiceáil 'sa Dúlraidh mórthimpeall air nó uathbhás do-eolais na síorraíochta a bhaint as an bhfláth feochta. Ba leor leis an fhile an radharc a chonnaic sé agus an glór a chuala sé gan aon mheabhrú a dhéanamh ortha. Ní leanann sé ó go raibh an Ceilteach dall ar mhórtas agus ar uaisleacht na Dúlraidhe agus nár thuig sé go raibh dia nó cruthaitheoir taobh thiar do'n iomlán. Níor dúil leis a leithéid a admháil nuair abhí an Nádúr dá canadh agus dá moladh aige. Bhí dlúth-charadas

> eadar an fhile agus an Nádúir, bhí sé ar aon-chéim léithi. Níor mhian leis an dá rud, an domhan corportha agus anam-cheist na síorraíochta mheascadh le chéile, agus is feairde a chuid filíochta nár mhian. (luaite ag de Paor 2018: 48)

Is ríluachmhar an t-eolas ar 'an domhan corportha' atá le fáil sa litríocht sin agus muid ag déileáil le géarchéim na haeráide. Agus é ag scríobh faoi Eochaidh Ó hEoghusa, duine de fhilí na scol, deir an file agus an scoláire Louis de Paor go dtagann an chuid is fearr den litríocht i gcónaí as 'teannas mar sin idir éileamh an traidisiúin agus brú na samhlaíochta aonair' (14). Is féidir a mhaíomh ar an taobh eile den scéal muna bhfuil aon éileamh ón traidisiún ann nach mbíonn an oiread sin brú ar an tsamhlaíocht. Teipeann uirthi an dochar atá á dhéanamh don timpeallacht a mhíniú nó fiú a shamhlú. Coincheap a bhí lárnach sa seanchreideamh mar gheall ar an nádúr ná go raibh 'an rí pósta ar bhandia na talún is go raibh cumhacht speisialta ag an bhfile mar idirghabhálaí idir an chríoch dhaonna a bhí faoi réir an taoisigh agus críocha nádúrtha agus osnádúrtha an bhandé' (23). Léiríonn anord sa nádúr anord sa chóras polaitiúil – 'Ar is gó flatha do-ber sína saíba for

tūatha clóena, co secca talman torad' (Thurneysen 1917: 82)[Tagann drochaimsir ar phobail ghránna ó fhalsacht an Phrionsa agus bíonn torthaí na talún seasc dá bharr]. Ní foláir teacht arís ar na hidirghabhálaithe – na scríbhneoirí, an t-aos eagna, lucht na cruthaíochta – chun an nasc idir an chríoch dhaonna agus na críocha nádúrtha a athchruthú agus a shlánú. Is léir ón traidisiún liteartha go raibh tuiscint pholaitiúil ag an aos eagna ar an dúlra. Bhí rath an phobail ag brath ar rath an nádúir. Ba chomhartha drochbhail an nádúir go raibh rud éigin as a riocht, go raibh géarchéim ann.

Tuigeann ealaíontóirí áirithe an gá atá le saibhreas cultúrtha na Gaeilge a shní isteach i saothar a chuireann síos ar thubaiste éiceolaíoch in Éirinn. Sa cheoldráma a scríobh agus a chum Donncha Dennehy faoin teideal *The Hunger* (2019) tugadh príomháit d'amhrán, 'Na Prátaí Dubha', a scríobh Máire Ní Dhroma ón Rinn aimsir an Drochshaoil. Tubaiste éiceolaíoch a bhí ann i gcás an Ghorta Mhóir mar bhí tromlach na ndaoine ag brath ar ghlasra amháin agus bhí toradh marfach ar mhonacultúr an bhia. Ach mar a léiríonn Ní Dhroma san amhrán ní raibh an tubaiste seo 'nádúrtha' in aon chor agus cáineann sí go géar na húdaráis idir thuata is chléir a bhí freagrach as cruachás na mbochtán sa tsochaí:

Na leanaí óga thógfaidís suas le macnas
Sciobtaí uathu iad gan trua gan taise dhóibh:
Ar bheagán lóin ach súp na hainnise
Gan máthair le freagairt díbh dá bhfaighidís bás.

A Rí na Trua is a Uain Ghil Bheannaithe,
Féach an ainnise atá dár gcrá
Is ná lig ar strae Uait Féin an t-anam bocht
Is a fheabhas a cheannaigh Tú é féin sa Pháis.

Nach trua móruaisle go bhfuil mórán coda acu
Ag íoc as an obair seo le Rí na nGrás;
Fearaibh bochta an tsaoil seo ná fuair riamh aon saibhreas
Ach ag síorobair dóibh ó aois go bás.

(Ní Dhroma 2019)

An dúshlán atá ann do ghlúin nua d'idirghabhálaithe ná go gcaithfidh siad an seanphósadh seo idir an chumhacht thuata agus an timpeallacht nádúrtha a cheiliúradh i mbruachbhailte, sna cathracha agus sna bailte móra, sna heastáit tithíochta. Cinnte, bíonn fonn ar lucht labhartha agus lucht scríofa na Gaeilge éalú ó ualach an mhiotais a cheanglaíonn an teanga i dtólamh leis an seansaol agus críocha sceirdiúla, uaigneacha na mbileog turasóireachta. Ach tá

teipthe ar eiseamláir na nua-aimsearthachta. An t-aon toradh a bheas ar fhás éigríochta ná léirscrios an domhain. Ní bhaineann traidisiúin na Gaeilge mar sin le haimsir chaite an mhaoithneachais ach le todhchaí réadúil na timpeallachta. Ní foláir dúinn an teannas cruthaíoch sin 'idir éileamh an traidisiúin agus brú na smahlaíochta aonair' a athlonnú i gcroílár na gcathracha móra mar aon leis na mbailte beaga tuaithe atá á dtréigean go tréan faoi láthair. Ach ní féidir é sin a dhéanmh gan an Ghaeilge a bheith i gcroílár an aonaigh. Soláthraíonn sí an t-eolas agus an stair agus na tuiscintí atá de dhíth orainn má táimid chun conartha nua nádúrtha a chumadh arís in Éirinn, athnuachan ar an gceangal ársa cleamhnais idir taoiseach is timpeallacht.

Ní féidir a shéanadh a thuilleadh go bhfuil chuile rud fite fuaite le chéile. Cruthaíonn, mar shampla, chuile mhála criospaí folamh fadhb bhreise do thodhchaí na timpeallachta in Éirinn mar nílimid in ann an plaisteach miotalach sna málaí a athchúrsáil (Hamilton 2019: 11). Ciallaíonn sé sin freisin go gcaithfimid agus muid ag dul i ngleic le fadhbbanna na géarchéime éiceolaíche tairbhe a bhaint as gach cuid den oidhreacht chultúrtha a bhaineann leis an Ghaeilge. Deir an Pápa Proinsias san Aitheasc Aspalta *Laudato Si* go bhfuil meas ar an timpeallacht ceangailte le feasacht aeistéitiúil:

> Má fhéachaimid ar an nádúr agus ar an timpeallacht gan a bheith oscailte don uafás agus don iontas, gan labhairt a thuilleadh i dteanga an bhráithreachais agus na háille inár gcaidreamh leis an gcruthaíocht is é an dearcadh a bheidh againn ná dearcadh na máistrí, an lucht tomhaltais, lucht an dúshaothraithe, gan trua ná taise, gan ar ár gcumas smacht a chur ar mhian na huaire. Ar an láimh eile, má bhraithimid aontaithe go dlúth le gach a bhfuil ann brúchtann stuaim agus cúram aníos uathu féin. (Proinsias 2016: 9)

Foinse amháin den oscailteacht atá á héileamh ag Proinsias ná traidisiún ársa i stair na Críostaíochta in Éirinn, is é sin, urraim don nádúr. Agus é ag scríobh faoi bhunaitheoir choincheap na hEorpa, Naomh Colmán, luann an staraí comhaimseartha Burnam Reynolds an bhaint a bhí ag Colmán leis na héanlaithe agus leis na hainmhithe fiáine, agus de réir an tseanchais, go raibh sé de nós ag scata de lachain fhiáine é a leanúint ó áit go háit agus an naomh i mbun a ghnó. Maíonn Reynolds, 'The Irish had long been noted for their love of nature and their conversion to Christianity did nothing to diminish that. It is a typically Irish custom to portray

their saints as having a special relationship with the created world. This is in stark contrast to much of the continental literature that viewed the world as "the devil's playground" and nature in general as something to be rejected.' (Reynolds 2012: 72).

Le cúpla bliain anuas, áfach, is léir go bhfuil na hÉireannaigh ag spraoi in áit súgartha an diabhail charbóin (O'Sullivan 2019) agus i bhfianaise dhíothú na bithéagsúlachta in Éirinn faoi láthair is beag den 'love of nature' sin atá le feiceáil inár bpolsaithe agus inár gcleachtais eacnamaíocha agus talmhaíochta (Fogarty 2017). Ach tá an dúil sin mar chuid dár n-oidhreacht agus ba chóir go mbeadh eolas fairsing ag muintir an oileáin ar an litríocht agus ar an tsaíocht a thug ómós ceart agus cruinn don domhan atá fós mórthimpeall orainn. Ar ndóigh, ní haon ionadh go bhfuil an dímheas ar an timpeallacht lárnach sa tsochaí thionsclaíoch, mar go dtí le déanaí ní raibh sna coillte agus sna haibhneacha agus sna slêibhte ach amhábhar do mhéadú na táirgíochta agus an bhrabúis. Mar a deir Proinsias, 'aon ní leochaileach mar an timpeallacht, níl aon chosaint aige ar chumhacht an mhargaidh a adhartar, geall leis, agus a ndéantar an t-aon riail de' (Proinsias 2016: 26). Tig linn cinnte an Ghaeilge a chur san áireamh mar 'ní leochaileach' a chuireann déistin ar lucht adhartha an mhargaidh uilechumhachtaigh, í ina constaic

chostasach ar bhealach an aonteagachais.

An loighic fhóntaíoch atá taobh thiar de léirscrios achmhainní nádúrtha an phláinéid, is follas go bhfuil sí ag feidhmiú freisin laistigh den díspeagadh ar an teanga mar ábhar 'úsáideach' nó scil 'phraiticiúil'. Ní féidir, mar sin, cúis na héiceolaíochta a chosaint agus neamhaird a thabhairt ar an nGaeilge. Baineann marthanas na timpeallachta le loighic fhrithfhóntaíoch a aithníonn tosaíochtaí sa saol seachas a bheith úsáideach, a thugann ómós d'ilghnéitheacht agus d'éagsúlacht. Braitheann teacht aniar an éiceachórais ar réimse leathan roghanna sa chóras ionas go mbeidh an córas in ann déileáil le fadhbanna nach raibh aon choinne aige leo. Conas is féidir sa chomhthéacs sin polasaithe a fhorbairt i leith na timpeallachta in Êirinn agus faic ann faoi ról na teanga? Mar níl aon teacht ar roghanna breise muna bhfuil a fhios ag éinne go bhfuil siad ann. Is é sin a rá go mbíonn chuile shochaí ag brath ar réimeas na hairde atá i gceannas inti. Tugann na saoránaigh suntas don rud seo nó don rud siúd agus ní hionann na hábhair suntais ó shochaí go sochaí, cúrsaí ciníochais sna Stáit Aontaithe, todhchaí na gcoillte san Fhionlainn, forbairt an deiscirt san Iodáil. Ní rud neodrach é, áfach, an aird. Caitheann comhlachtaí móra an t-uafás airgid ag iarraidh ár n-aird a dhíriú ar earraí nó ar sheirbhísí faoi leith ar an idirlíon nó

trí na meáin shóisialta (Beck and Davenport 2001).

Níl aon earra níos luachmhaire sa gheilleagar comhaimseartha ná ár n-aird. Tuigeann polaiteoirí fresin cé chomh tábhachtach is atá sé aird an phobail a tharraingt ar ghné amháin de scéal agus gan aon aird a dhíriú ar ghné eile, pobalóirí ag tabhairt amach faoi ualach na n-imirceoirí ar an gcóras sláinte agus iad ina dtost ar fad i dtaobh ollchuidiú eacnamaíoch na n-imirceoirí céanna. Baineann réimeas airde go dlúth le teanga. Tugaim suntas do rudaí áirithe trí na hacmhainní a chuireann an teanga ar fáil dom. Is tríd an nGaeilge a fhaightear léamh iomlán éagsúil ar chruth agus ar stair na tíre. Cuireann an teanga faoi do bhráid an iliomad gnéithe de thimpeallacht nádúrtha agus daonna an oileáin nach bhfuil ar fáil sa Bhéarla. Agus an ghéarchéim aeráide ag teannadh linn ní féidir linn a thuilleadh bheith dall ar chumas múscailte na teanga. Níl an dara rogha againn nó ní thiocfaidh ár lá go deo na díleann.

Inbhuanaitheacht

Ní raibh amhras ar bith ar an Athair Peadar Ó Laoghaire agus é ag scríobh eagarfhocail don iris *Fáinne an Lae* in 1898 ná go raibh práinn intleachtúil agus mhorálta le teanga a chothú:

> Do choiméad an t-urlabhra sain os comhair na n-aigneadh eolas ar sheanchas na sinsear reompa; eolus ar dheighthréithibh na sinsear sain; ar a n-urraim do gach maith, ar a ngráin ar gach olc; ar a ndúil i gceart, ar a bhfuath do'n eugcóir; ar uaisleacht a n-aigneadh ar mhéidh a gciall. Do dheintí, mar ba dhual, aithris ar na deighthréithibh sin. Do tugthaoi urraim do'n aois agus do glacthaoi an chomhairle thigeadh uaithi, agus do dheineadh gach sliocht díchioll ar theacht suas le gach sinsear, i leannmhaint na ndeighbhéas, agus i dtuilleamh na ndeaghchlú.

Cuireadh an bhéim ar an gceangal idir an teanga agus an tsinsearacht, idir lucht labhartha na Gaeilge agus Ré Órga na nDea-Bhéas. Ach cad faoin ngaol idir teanga agus todhchaí, idir urlabhra agus inbhuanaitheacht? Agus é ar an mbád go Sasana

in 1981 feiceann sagart eile, an scríbhneoir agus an taistealaí Frainc Mac Brádaigh, bean óg ag gol. Dúisíonn íomhá truamhéalach seo na heisimirce fearg an tráchtaire shóisialta, "Dá dtabharfadh muid aird ar na héagóracha is an t-athmhachnamh is gá a dhéanamh, thuigfeadh muid go bhfuil athruithe bunúsacha de dhíth leis an tsochaí" (Mac Brádaigh 2010: 9). Bíonn macalla na bhfocal sin le cloisteáil arís agus muid sáite i ngéarchéim éiceolaíoch atá ar bhealach níos measa agus níos baolaí ná an ceann a ghríosaigh an bhean chroíbhriste sin chun siúil ag tús na n-ochtóidí. Ní foláir athmhachnamh iomlán a dhéanamh ní hamháin ar na héagóra bunúsacha sa tsochaí ach ar ról na teanga i bhforbairt agus in athmhúscailt an oileáin. Tá dánaíocht na smaointeoireachta réabhlóidí de dhíth orainn arís ach cén cruth a bheas ar an léirsmaoineamh radacach seo?

De ghnáth cuirtear síos ar an ré chomhaimseartha mar ré an domhandaithe. Tharla athruithe bunúsacha sa chóras geilleagrach idirnáisiúnta ag deireadh na haoise seo caite a chuir tús le réabhlóid nua i gcúrsaí eacnamaíochta. Ní raibh an córas Fórdach a bhí bunaithe ar olltáirgeadh agus ar fhorlámhas na gcomhlachtaí móra ag feidhmiú a thuilleadh. D'fhás an táirgeadh iar-Fhórdach as an dlúthcheangal idir cumhacht na ngréasan ríomhaireachta agus

teileachumarsáide agus córas táirgthe iomlán díláraithe. Bhí comhlachtaí in ann earraí a dhéanamh i monarchain scaipthe ar fud an domhain ar chostas an-íseal, agus le cabhair na teicneolaíochta faisnéise ní raibh deacracht ar bith le táirgeadh agus soláthar na n-earraí seo a stiúradh. Cuireadh deireadh le rialacha maidir la malartú airgid, agus ba é an bhunsprioc a bhí ag an gComhaontú Ginearálta do Tharaifí agus Trádáil agus ag an Eagraíocht Trádála Domhanda ná fáil réidh le bacainní taraifí. Tháinig fás ollmhór ar thionscal na fógraíochta a bhí dírithe go príomha ar mhianta luaineacha na dtomhaltóirí iar-nua-aimseartha ar leibhéal domhanda (tábhacht an 'bhranda'). Is féidir a rá gurbh ionann réabhlóid an domhandaithe agus réabhlóid an tseachtraithe. Is é sin a rá gur thréith shuntasach den chóras nua an seachtrú, go raibh comhlachtaí in ann chuile fheidhm, beagnach, de chuid na comhlachta a lonnú in áit ar bith. Ar ndóigh, bhain Éire an-leas as an gcor nua seo i scéal an gheilleagair.

Ní féidir, áfach, leathnú gan stad i spás críochta. Ní féidir ach an oiread baint i gcónaí ó acmhainn chríochta. Tuigtear anois nach bhfuil ach domhan amháin againn agus acmhainní gann agus teoranta ar an bpláinéad sin. Má leantar ar aghaidh leis an eiseamláir gheilleagrach atá i bhfeidhm faoi láthair scriosfar an timpeallacht atá mar bhunús

riachtanach do mharthanas an chine dhaonna. Ní mór athmhacnamh a dhéanamh ar chuile ghné d'fheidhmiú na sochaí ionas go mbeidh seans ag saoránaigh an domhain na deacrachtaí dosheachanta seo a shárú. Séard atá ag teastáil anois ná réabhlóid nua, is é sin *réabhlóid an inmheánaithe*. Ciallaíonn réabhlóid an inmheánaithe go mbeidh todhchaí agus inbhuanaitheacht na sochaí as seo amach ag brath go tréan ar an gceangal leis an áit logánta. Sa bhliain 2005 bunaíodh gluaiseacht i gCionn tSáile i gCorcaigh chun tús a chur leis an réabhlóid. I gcomhthéacs buaic ola agus athrú aeráide an sprioc atá ag *Transition* nó Trasdul ná úsáid fuinnimh agus acmhainní teoranta eile a laghdú go suntasach. I ngach aon réimse – bia, iompar, fuinneamh teaghlaigh – cuirtear an bhéim ar earraí agus seirbhísí a fhoinsiú san áit logánta féin. Déantar é sin trí gharraithe pobail, córais bhabhtála, ceardlanna athchúrsála agus araile a eagrú. Mar chuid den réabhlóid riachtanach seo ní foláir athcheangal a dhéanamh arís leis an áit logánta, ach chun é sin a dhéanamh ní mór dúinn teacht ar thuiscint ar nádúr, ar stair agus ar thualaingt na háite. Bunchloch na tuisceana sin sa tír seo ná an Ghaeilge a thugann eolas as cuimse dúinn madir le seanchas, logainmneacha, dinnseanchas, miotaseolaíocht, amhráin a bhaineann le háit ar bith ar an oileán. Dá bhíthin sin, ba chóir go mbeidh an teanga i gcroílár chuile iarracht chun

inbhuanaitheacht na sochaí a chinntiú sna blianta atá romhainn.

Is féidir a mhaíomh gurbh í an réabhlóid ba thábhachtaí i stair na hÉireann ag deireadh an naoú haois déag ná go bhfuair muintir na hÉireann (nó cuid díobh ar aon nós) athsheilbh ar a dtalamh féin. An bhunaidhm a bhí ag lucht Conradh na Talún ná fáil réidh le héagóir an dísheallbhaithe choilínigh. Mar thoradh ar an ngluaiseacht sin, áfach – agus ní féidir fiúntas ná cóir na gluaiseachta féin a shéanadh – tháinig cultúr nua i dtír a fuair greim docht ar mheon na nÉireannach, is é sin, cultúr na seilbhe. Le linn bhlianta an Tíogair Cheiltigh ba léir go raibh reiligiúin seo na réadmhaoine i mbarr a réime agus luach na ditithe agus na talún de ló is d'oíche i mbéal an phobail. Tá an tír anois breactha le heastaít tithíochta scáinte, mar mheafar truamhéalach ar chultúr santach na seilbhe. Sin an dara réabhlóid atá ag teastáil ná bogadh ó chultúr na seilbhe go *cultúr an chónaithe*. I dtéarmaí eile, is tábhachtaí ar fad timpeallacht ná sealúchas. Caithfimid freastal ceart, íogair a dhéanamh ar an áit ina bhfuilimid in ionad a bheith scartha amach uaithi ar árthach spáis an ábharachais. Ar bhealach, níl an dara rogha againn, muna n-aithnítear cinnireacht bhunúsach na timpeallachta beidh deireadh leis an saol mar atá. Sa chomhthéacs seo, conas is féidir le héinne bheith ag

maireachtáil go tuisceanach agus go leochaileach in aon áit ar oileán na hÉireann gan cur amach aige nó aici ar an teanga a chuimsíonn an oiread sin eolais ar an tír máguaird? Ar ndóigh, cuireadh tús le gluaiseacht na hAthbheochana aimisir Chogadh na Talún. Anois agus réabhlóid eile de dhíth orainn chun dul i ngleic le fadhbanna práinneacha an athraithe aeráide, tá ról lárnach arís ag an teanga chun fíorchultúr an chónaithe a chruthú, a chothú agus a spreagadh.

Thabharfadh an dá réabhlóid seo, réabhlóid an inmheánaithe agus réabhlóid an chónaithe, aitheantas ceart don teanga dhúchais ach ní mar rud dúnta, seasc, seachantach ach mar eochair na fuascailte, doras an feasa, gléas na fairsinge. Thuig Séamus Ó Doireadháin san ochtú haois déag an méid sin nuair a chum sé an dán úd in ómós do Sliabh Liag.

Is ag sáil Chnocán Áine tá cúirt gheal is parlús
Aolbhallaí bána agus táiplis 'na suí
Is go mórbhruínte an tsléibhe a ghluaiseas na céadta
'Un bainse 'gus 'un féasta is a leithéid as gach tír
Bídh slua de mhná Gréagach' dualach' dea-éadaigh
Is cláirseach ar théadaí a mbréagadh agus píob

D'oscail focail Uí Dhoireadháin cúirt gheal na timpeallachta. Ba chóir anois go mbeidh ceolchoirm sin na héagsúlachta le cloisteáil i ngach aon cheard den oileán.

Ní nach ionadh i ré seo an domhandaithe, luaitear go minc coincheap na saoránta chta domhanda. Chun teacht i dtír ar na fadhbanna éagsúla a dhéanann an oiread sin dochair don timpeallacht ní foláir féachaint orainn féin mar shlabhraí leochaileacha den phobal daonna. Ach an é sin an bealach ceart chun ár ndeacrachtaí a shárú? Cad is brí le saoránach ach duine atá in ann fáil réidh le laincisí na háitiúlachta, atá ábalta lorg an dúchais a chaitheamh uaidh nó uaithi chun dul i dtreo an uilíochais? Mhaígh Arastatal gurbh ionann saoirse agus spleáchas. Is é sin a rá gurbh é an difríocht ba mhó idir an sclábhaí agus an duine saor ná nach raibh ceagal ar bith ar an sclábhaí. Bhí de cheart ag an máistir é a bhogadh ó áit go háit ar nós earra. Ní mar sin a shamhlaítí an duine saor, áfach, toisc go raibh gaolta aige agus go raibh sé ceangailte le háit faoi leith. Bhí a chuid saoirse bunaithe ar an idirspleáchas seo idir é féin agus a mhuintir agus a áit dúchais. Sa chomhthéacs seo ba chóir coincheap níos sásúla a aimsiú do ghné spleách an chine dhaonna. I Sasana tá éagraíocht ann darb ainm *Common Ground* (www.commonground.org.uk) agus an príomhchúram atá ar an eagraíocht

ná cultúr, seanchas, agus miotaseolaíocht áiteacha éagsúla ar fud na tíre a chaomhnú agus a cheiliúradh.

Úsáidtear lucht *Comnon Ground* an téarma taithítheoir ((*denizen*) chun cur síos ar an té a mhaireann agus a ghníomhaíonn go feasach in áit faoi leith, is cuma más áit tuaithe nó áit chathrach í. Murab ionann is an saoránach, níl an taithítheoir scartha amach ón áit nó ón bpobal dúchais. Aithnítear spleáchas an duine ar a thimpeallacht nádúrtha, idir ábharaíoch agus dhaonna. Má ghlacaimid leis go mba chóir bogadh ó chultúr an tsaoránaigh go cultúr an taithítheora, caithfear a admháil freisin go bhfuil feasacht agus inbhuanaitheacht an taithítheora fite fuaite le tuiscint fhairsing chruinn ar an eolas a sholáthraíonn an teanga dhúchais ar chuile ghné de choimíneacht na hÉireann.

Ní éilíonn an tríú hathrú réabhlóideach seo, ámh, bogadh i dtreo chultúr an taithítheora agus ceiliúradh saonta a dhéanamh ar shuáilcí na haontachta nó ar dheasghnátha an chúigeachais. Is léir ní hamháin ó fhianaise na staire sóisialta (féach, mar shampla, leabhar Mhichíl Uí Chonghaile, *Conamara agus Árainn 1880-1980: Gnéithe den Stair Shóisialta* (1988)) ach ón gcur síos géar atá déanta ag an-chuid scríbhneoiri ó Phádraig Ó Conaire agus Máirtín Ó Cadhain go Diarmuid Ó Súilleabháin agus Eilís Ní Dhuibhne go mbíonn teannas agus easaontacht agus

síorchoimhlint ann in áit ar bith. In úrscéal Lorcáin S. Uí Threasaigh *Cnoc na Lobhar* (2007) tuigeann an príomhcharachtar Labhrás nach féidir coincheap na muintire, mar shampla, a scaradh amach ó aighneas inmheánach, do-sheachanta an ghrúpa. Agus é sínte ar an tolg tar éis oíche mhór ólacháin ní mór dó a admháil nach suaimhneas i gcónaí a bhíonn le fáil i measc do mhuintire:

> Seo é an teach agam, seo iad mo mhuintir agus seo é mo shaol, a bhí mé ag ceapadh sular thit mo chodladh orm sa tolg agus mé ag brionglóidí ar a bheith in áit éigin eile ar thaobh eile an domhain seachas a bheith i m'aonar i ngarraí leathscoite i mo shaoránach i measc na saoránach leathscoite ag casadh na cré ar mhaithe lena casadh go dté mé lá éigin inti, cré dhubh mo mhéine, cré fhuar na cille. (142)

Ciallaíonn cultúr an taithítheora glacadh leis an achrann a bhaineann le pobal ar bith agus tarraingt as acmhainní dúchasacha an chultúir chun deacrachtaí daonna agus timpeallachta a shárú. Sa deireadh thiar thall, níl ciall le réabhlóid an inmheánaithe ná le réabhlóid an chónaithe muna bhfuil figiúr dúshlánach an taithítheora ghrinnsúiligh i gcroílár an aonaigh.

Ag tús an chéid seo ba léir go raibh rogha ag Éire dul bealach amháin nó bealach eile. Bealach amháin ná dul i dtreo forbairt a bheadh bunaithe go príomha ar nuálacht, ar eolas, ar acmhainní cultúrtha na sochaí Éireannaí. An bealach eile ná forbairt bunaithe ar earnáil na tógála, ar impireacht na mbrící agus ar theacht isteach ioncaim ó dhleacht stampála (O'Toole 2009). Rognaíodh an dara bealach, faraor, agus tar éis 2008 bhí an tír báite in umar na haimléise. Thaispeáin an tréimhse sin i stair na tíre an baol a bhaineann le monachultúr. Má bhíonn geilleagar nó sochaí ag brath ar aon chur chuige amháin, ar aon léamh amháin ar dhul chun cinn na tíre, bíonn leochaileacht de shíor mar dhlúthchuid de thodhchaí na sochaí. Bíonn blas ar an mbeagán agus is fíorshearbhasach an blas go fadtéarmach muna mbíonn ann ach an beagán. Thaispeáin cruachás na hÉireann go suntasach tábhacht phrionsabal na hathléimneachta, prionsabal atá bainteach leis na hathruithe réabhlóideacha a luamar cheana. Is ionann prionsabal na hathléimneachta agus béim a chur ar éagsúlacht a chruthú agus a chaomhnú i sochaí, éagsúlacht cultúir, éagsúlacht teanga, éagsúlacht creidimh, éagsúlacht luachanna. I gcás géarchéime, is féidir dul chuig an iliomad tobar eagna seachas a bheith i dtólamh ag tarraingt ón tobar aonair, spleáchas cultúrtha a chothaíonn éiginnte agus

leochaileacht i measc an phobail.

Conas prionsabal na hathléimneachta a chur i bhfeidhm in Éirinn? Céim amháin ná dlús a chur le bonneagar intleachtúil na tíre a thógáil in ionad a bheith de shíor ag cur béime ar an mbonneagar fisiciúil. Ciallaíonn bonneagar intleachtúil tionscnaimh nó gréasáin a fhorbairt nó a mhaoiniú a chothaíonn tóir an eolais, fás na cruthaíochta agus scaipeadh na smaointeoireachta, bunchlocha na héagsúlachta in aon tír faoi leith. Ar ndóigh, deirtear go minic gurb í sin an sprioc atá ag 'sochaí an eolais' nó ag an 'geilleagar cliste' ach de ghnáth ní chuirtear san áireamh ach an eolaíocht agus an teicneolaíocht agus na téarmaí seo á lua ag polaiteoirí agus ag lucht na meán, agus déantar neamhshuim de chuile ghné eile de chultúr na sochaí. Fiú na maithe agus móruaisle a tháinig le chéile i dTeach Farmleigh in 2010, d'aithin siad go raibh fiúntas faoi leith ag baint le saibhreas cultúrtha na tíre agus b'amaideach gan tairbhe a bhaint as an stór luachmhar sin. Luíonn sé le réasún mar sin go mbeadh ról lárnach ag an nGaeilge agus an bonneagar intleachtúil seo á fhorbairt againn. Ach, ar ndóigh, ní mar a shíltear a bhítear. Nuair a chuir an *Irish Times* sraith alt i gcló cúpla bliain ó shin mar gheall ar Athnuachan na Poblachta, ní raibh oiread is alt amháin ann faoin teanga agus an cultúr a bhaineann léi. I ré na hAthbheochana

agus an t-aos eagna ag cumadh friotal úr polaitiúil i gcomhair na poblachta nua, bhíodh ceist na teanga i gcroílár na ndíospóireachtaí san am. Anois, ní raibh oiread is paragraf amháin ann faoin teanga agus an Phoblacht Nua idir chamáin ag ár smaointeoirí agus ag ár scríbhneoirí dúchasacha.

Ach cá bhfuil an fhianaise go bhfuil teacht ar shúil eile, ar éagsúlacht, ar nuaíocht i saíocht na Gaeilge? Is féidir neart samplaí a lua ach ba mhaith liom anseo cloí le réimse an léinn. I leabhar Bhriain Uí Chonchubhair *Fin de Siècle na Gaeilge* (2009) tugtar léargas fíorspéisiúil dúinn ní hamháin ar an gcomhthéacs Eorpach d'athbheochan na teanga ach tá scagadh an-tráthúil ann freisin ar fhorbairt ionannais áitiúil i sochaí a bhí (agus atá fós) de shíor faoi bhrú ag fórsaí domhanda theanga na hImpireachta. Déanann Caoimhe Máirtín sa leabhar *An Máistir: an scoil agus an scolaíocht i litríocht na Gaeilge* (2003) cur síos spreagúil ar na claochluithe oideachais in Éirinn agus conas mar a chuaigh siad i bhfeidhm, go mór mór ar chosmhuintir na tíre. Ardaítear ceisteanna bunúsacha maidir le bunús fhealsúnacht an Oirthir i leabhar Eoghain Mhic Aogáin, *Johannes ó Éirinn agus Máistrí na Gréige* (2009) agus tugann Helen Ó Murchú lón machnaimh don léitheoir faoi chúrsaí creidimh agus í ag trácht ar stair na Críostaíochta in Éirinn in *An Chríostaíocht le Sinsearacht* (2010).

Níl anseo ach sampla beag bídeach de na saothair bhreátha, théagartha a thagann amach go rialta sa Ghaeilge, ach cá bhfuil teacht ar léirmheasanna ar na leabhair seo, fiú sna meáin chumarsáide Gaeilge? Cá bhfuil na cláir leabhair ar an raidió nó ar an teilifís a scaipfeadh dea-scéal na hintinne dúchasaí? Conas is féidir aon cheo a rá mar gheall ar shocaí an eolais, ar shocaí chliste, muna bhfuil teacht ar na hacmhainní eagna seo? Níl ciall ar bith le bonneagar intleachtúil a chruthú má fhágtar bunchloch na teanga agus an tsaíocht atá taobh thiar di ar lár. Má táimid i ndáiríre faoi Phoblacht a athnuachan, faoi shochaí inbhuanaitheach a chruthú do na glúnta atá romhainn, níl beart níos fearr dúinn ná preab a chur arís i ngníomh ceannairceach na teanga.

Eolas

Timpeall na bliana 1657 chum file darbh ainm Seán Ó Conaill dán dar teideal, 'Tuireamh na hÉireann'. Níl an dán ar eolas ag mórán daoine anois ach bhí tráth agus de réir Vincent Morley ba dheachair neamhshuim a dhéanamh de théacs a bhí chomh lárnach sin i bhforbairt ionannas an mhóraimh in Éirinn:

> Más aon teist iad an líon mór lámhscríbhinní ina bhfuil an téacs ar fáil, is féidir a rá go raibh ráchairt as cuimse ar "Tuireamh na hÉireann" ar feadh dhá chéad bliain nach mór tar éis a chumtha. Ba é an dán seo, thar aon saothar eile, a mhúnlaigh tuiscint na nGael ar stair na hÉireann agus a chuir insint Chaitliceach de stair na tíre ar fáil do chosmhuintir nach raibh leabhair, ná léamh, ná Béarla acu. Caithfear aiste Sheáin Uí Chonaill a áireamh mar cheann de na téacsaí ba thábhachtaí dár scríobhadh in Éirinn riamh. (Morley 2011: 68)

Bhí dán Uí Chonaill go mór faoi anáil mhórshaothar an Chéitinnigh, *Foras Feasa ar Éirinn*, ní nach ionadh,

mar bhí an saothar céanna ina fhoinse údarásach ag údair uaslathacha an tseachtú haois déag a bhí faoi thionchar fhilíocht na scol, ag filí a chum dánta sa dá theanga agus a chleacht meadarachtaí an amhráin san ochtú haois déag , agus ag filí pobail an naoú haois déag ar mhair dánta leo sa bhéaloideas go dtí tús ré na athbheochana. D'éirigh le Céitinn léann traidisiúnta na Gaeilge agus léann comhaimseartha na hEorpa barócaí a bhí sealbhaithe aige a nascadh le chéile agus cruth nua-aoiseach a chur ar sheanchas na hÉireann. Murab ionann is na hAnnála a scríobhadh i dteanga bhréagársa agus a bhí in oiriúint don aos dána, dream a fuair a gcuid oideachais sna bardscoileanna nach raibh ann a thuilleadh, bhí cuntas leanúnach, soléite le fáil i stair an Chéitinnigh a thug lón intinne agus ábhar dóchais do lucht an mhóraimh ar feadh tréimhse fada.

Cá bhfuil leithéidí Shéain Uí Chonaill nó Sheathrúin Chéitinn anois? Cá bhfuil an aiste a chothódh misneach agus éirim aigne ní hamháin i measc lucht labhartha na Gaeilge ach i measc mhuintir na hÉireann i gcoitinne? An bhfuil sé d'inmhe ionainn cuntas nua, agus an ghéarchéim éiceolaíoch ag bagairt orainn, a sholáthar do phobal nua? An é 'Tuireamh' nó 'Turgnamh' na hÉireann atá romhainn amach anseo? Ba mhaith liom eilimintí a ainmniú a bheadh mar chuid den bhunábhar sa

turgnamh iar-nua-aoiseach seo.

An chéad eilimint ná fáil réidh le bréagdhearcadh ar stair na hÉireann agus ar chultúr agus ar thaithí dhaonna an mhóraimh le linn na gcéadta bliain. Is é sin a rá go bhfuil an dioscúrsa faoin teanga sa tír seo go mór faoi anáil mheafar bitheolaíoch an bháis. An téarma atá cumtha agamsa chun cur síos ar an dearcadh seo ná bunúsaíochas teileolaíochta. De bharr an Ghorta Mhóir agus thionchar tubaisteach an drochshaoil ar líon na gcainteoirí Gaeilge agus forbairt an spáis phoiblí sa teanga úsáidtear *telos* ('críoch' nó 'sprioc' sa tSean-Ghréigis) an scriosta nó an mheatha chun cur síos ar stair iomlán na teanga. Sa fhráma tagartha seo, níl ann sa Ghaeilge ón séú haois déag ar aghaidh ach sórt réamhléiriú feistithe i gcomhair dhráma fuilteach an deiridh. Ach mar a dúirt Breandán Ó Buachalla in *Aisling Ghéar* (1996) agus é ag cur síos ar scoláireacht na bProinsiasach Éireannach i Lováin, ní raibh na scoláirí ag tógáil leac cuimhneacháin bhláfar, bhladhmannach ar shibhialtacht a bhí ag saothrú an bháis ach iad gafa i dtionscnamh stairiúil nua-aoiseach a ghlac leis go raibh géarghá le hathnuachan intleachtúil chun cruth ceart a chur ar theanga nua-aimseartha ar bith.

Ag tús an naoú haois deag, ní haon 'moribund tradition' mar a mhaíonn Joep Leerssen in *Mere Irish and Fíor-Ghael* (1996: 315), a bhí i gcultúr na

nGael ach a mhalairt. Bhí níos mó éilimh ná riamh ar chóipeanna de 'Tuireamh na hÉireann'. Mar a deir Morley, 'Bhí aistí úra á gcumadh sna blianta sin ag Tomás Rua Ó Súilleabháin, Diarmaid na Bolgaí Ó Sé, Seán Ó Braonáin, Peatsaí Ó Callanáin, Mícheál Óg Ó Longáin and Eoghan Ó Comhraí agus tuilleadh nach iad' (Morley 2011: 264). Ar ndóigh, níl aon ghantannas samplaí ann, ach tá fíorghantannas ann maidir le tuiscint. Ról amháin atá ag an lucht léinn sa tír seo a bhfuil scileanna sa teanga acu ná dul i ngleic leis an mbunúsaíochas teileolaíoch marfach seo a chothaíonn an oiread sin aineolais i measc an phobail.

Dúirt Saoi na Teilifíse Bernice Harrison san *Irish Times* agus í ag caint faoi chlár ar TG4, 'most of us can hardly remember more than a cúpla focal of what was so badly taught at school. And we couldn't be bothered anyway' (Harrison 2012: 8). Is furasta an t-aineolas bagrach atá taobh thiar de na tuairimí sin a cháineadh, ach ní leor sin. Caithfimid iad a bhréagnú ní hamháin trí mheán na scoláireachta agus trí mheán na litríochta agus tá mé ag smaoineamh anseo ar na húrscéalta stairiúla, bríomhara, tuisceanacha, spreagúla a tháinig amach le cúpla bliain anuas, leithéidí *An Litir* (2012) le Liam Mac Cóil, *Scéal Ghearóid Iarla* (2011) le Máire Mhac an tSaoi nó *An Cléireach* (2008) le Darach Ó Scolaí, ach fós féin

trí mheán na músaem, na ndánlann, na leabharlann agus le cabhair na meán uilig. Ní leor a thuilleadh gort an eolais a threabhadh. Caithfimid torthaí an fhómhair a chur ar taispeáint i gcroílár an aonaigh chumarsáide. I sochaí an eolais ní fiú faic an t-eolas é féin. Tagann an luach as scaipeadh an eolais. Ní foláir comhthéacs a chruthú ina mbeadh scoláirí, scríbhneorí, ealaíontóirí, saoithe teanga go leanúnach ag bréagnú bholscaireacht na baoise a bhíonn mar bhunchloch ag tráchtairí áirithe agus iad ag tabhairt breithiúnais ar stair, staid agus todhchaí na Gaeilge.

An dara heilimint atá le cur san áireamh agus muid ag obair ar an turgnamh iar-nua-aoiseach seo ná nádúr nó feidhmeanna na teanga féin. I leabhar a tháinig amach i 2004, *L'Avenir des languages: repenser les humanités* déanann beirt scoláirí, Heinz Wismann agus Pierre Judet de la Combe, idirdhealú idir teanga na seirbhíse (*langue de service*) agus teanga an chultúir. Is ionann teanga na seirbhíse agus teanga a úsáidtear chun eolas ionstraimeach a chur ó chainteoir amháin go cainteoir eile. Is ionann teanga an chultúir agus an teanga a úsáidtear chun ár gcuid mothúchán, ár gcuid smaointe fánacha, ár gcuid samhlaíochta a chur faoi bhráid ár lucht éisteachta. Is léir, mar shampla, go mbíonn an Béarla ag feidhmiú go rialta mar theanga sheirbhíse ag cruinnithe agus ag comhdhálacha idirnáisiúnta. Is fíor freisin go raibh an-chuid den

agóidíocht ar son na Gaeilge le blianta beaga anuas dírithe ar an nGaeilge mar theanga sheirbhíse mar go raibh géarghá le cearta cuí le go mbeadh deis ag lucht labhartha na teanga a ngnó a dhéanamh trí Ghaeilge le ranna Stáit agus eagrais nach iad. Ach is baolach nach leor an dearcadh ionstraimeach sin agus muid ag iarraidh ról lárnach a aimsiú don teanga inár sochaí féin. Mar ní mór díriú ar cad is brí le leithleachas teanga.

Tig linn a mhaíomh go bhfuil trí ghné a bhaineann leis an leithleachas seo. Ar an gcéad dul síos, is ionann teanga faoi leith agus tacar struchtúr teanga faoi leith, agus cinnte is ábhar díospóireachta é an tionchar a bhíonn ag na struchtúir sin ar ár meon mar phobal. Baineann an dara gné leis na cleachtais shóisialta a bhíonn ag an bpobal teanga agus a chuirtear in iúl trí mheán na teanga. Maidir leis an tríú gné is léir gur pobal dioscúrsa nó ba chirte a rá tacar pobal díoscúrsa é chuile phobal teanga. Léiríonn na pobail dhioscúrsa seo stair, taithí dhaonna agus córais chreidimh lucht labhartha na teanga trí mheán na ndioscúrsaí atá forbartha laistigh den teanga. Baineann leithleachas stairiúil agus dlús séimeantach lucht labhartha teanga lena réimsí tagartha cainte, idir shiombalach agus chomhchoiteann. Is é, is dóigh liom, an dara cúram atá orainn ná díriú ar an teanga chultúir, gan dearmad a dhéanamh, cinnte, ar an teanga sheirbhíse, chun gur

follas go bhfuil réimsí tagartha faoi leith a bhaineann leis an teanga. Muna bhfuil, ní fiú í a labhairt mar go bhfuil an Béarla mar theanga sheirbhíse i bhfad níos cumhachtaí ná an Ghaeilge. Is é sin le rá go bhfuil cosaint na Gaeilge mar theanga sheirbhíse sa Ghaeltacht agus lasmuigh di bunaithe ar luachanna atá fite fuaite leis na gnéithe atá luaite agam. Ní rud imeallach nó uaslathach mar sin an teanga chultúir a shainmhíniú. Ba chóir go mbeadh an gníomh sin, is é sin, leithleachas cainte, leithleachas cleachtais agus leithleachas dioscúrsa a mhíniú go seasta agus go leanúanch, mar dhlúthchuid d'aon phlean gníomhaíochta as seo amach agus mar chuid den fheachtas éiceolaíoch laistigh den teanga féin.

An tríú heilimint agus muid ar thóir an chuntais nua ná dul i ngleic le hidéolaíocht an chostais. Cad is brí leis an idéolaíocht seo? Ciallaíonn sí go beacht gur rud neodrach é coincheap an chostais. Ní féidir linn an rud seo nó an rud siúd a dhéanamh mar tá sé róchostasach. Níl an t-airgead againn. Is oth liom a rá, tá brón orm, níl mise atá freagrach as seo, an dtuigeann tú, ach níl pingin rua fágtha againn agus caithfimid na costais a laghdú ar bhealach amháin nó ar bhealach eile. Féach an méid airgid a chaitear ar mhúineadh na Gaeilge sa chóras scolaíochta. Féach an méid airgid a chaitear ar aistriú tuarascálacha bliantúla. Féach an méid airgid a chaitear ar TG4 nó

ar Fhoras na Gaeilge nó ar Údarás na Gaeltachta. Baintear úsáid as coincheap an chostais mar shórt slat tomhais neamhspleách a dhíríonn aird an phobail ar réimsí áirithe nach féidir a chosaint toisc go n-alpann siad an iomarca airgid ó chiste an phobail, airgead nach bhfuil ann a thuilleadh. Ach cad tá taobh thiar den chaint leanúnach seo mar gheall ar chostais? I leabhar a tháinig amach bliain roimh a bhás scríobh an staraí agus an tráchtaire cultúrtha Tony Judt an méid seo a leanas:

> For thirty years we have made a virtue out of the pursuit of material self-interest: indeed, this very pursuit now constitutes whatever remains of our sense of collective purpose. We know what things cost but we have no idea what they are worth. (Judt 2010: 1)

Tuigeann Judt go bhfuil ceangal ann idir luach agus costas, go bhfuil dioscúrsa an chostais mar ghléas troda eile i gcomhrac inmheánach na sochaí agus grúpaí áirithe ag iarraidh a bhfís féin a bhrú ar an bpobal i gcoitinne. Smaoinigh, mar shampla, ar an argóint seo mar gheall ar chostais an aistriúcháin. Deirtear go minic gur náireach an rud é go bhfuil Rialtas na hÉireann nó an tAontas Eorpach ag caitheamh an

dúrud airgid ar chúrsaí aistriúcháin. Nach cóir an t-airgead sin a chaitheamh ar ospidéil nua nó ar scoileanna nó ar sholáthar córais leathanbhanda i gceantair iargúlta? Nach mbeadh sé níos fearr agus níos saoire an teanga chéanna a bheith ar eolas ag chuile dhuine gan a bheith ag brath ar threalamh costasach, ollmhór an aistrithe? Ach ceileann na ceisteanna seo ceist eile mar gheall ar chostas. Is é sin nach laghdú costais atá i gceist ach aistriú costais.

Mar a thaispeánann an teangeolaí Ceanadach François Grin, caitheann rialtais ar fud an domhain na céadta milliún dollar ag múineadh Béarla dá gcuid saoránach (2004: 189-202). Níl na Béarlóirí ag íoc as seo, ar ndóigh, ach na pobail agus na náisiúin nach labhraíonn an teanga áirithe sin. Ar go leor bealaí, bíonn na costais a bhaineann le múineadh teanga domhanda i bhfad níos airde ná na costais a bhaineann le cleachtadh an aistriúcháin ach is féidir a rá i gcás amháin go bhfuil na fíorchostais ceilte ach sa chás eile go bhfuil reitric an chostais á húsáid chun éagsúlacht teanga agus cultúir a scriosadh.

Go bunúsach, ní mór a aithint go bhfuil ceist an chostais fite fuaite le gné faoi leith de fheidhmiú an mhargaidh i láthair na huaire. Glactar leis anois go bhfuil athrú sainiúil tagtha ar ról an tomhaltóra sa tsochaí chomhaimseartha. Cad a tharlaíonn nuair a cheannaíonn tú ticéad eitilte Ryanair? Caithfidh tú

do phas bordála a chlóbhualadh roimh ré. Éilíonn sé sin go bhfuil teacht agat ar an trealamh cuí (ríomhaire agus printéir) agus go bhfuil nasc idirlín agat. Caithfidh an paisinéir nó duine éigin eile íoc as an trealamh sin agus as an nasc. Deischostas eile ná an t-am a chaitear ag iarraidh teacht ar an suíomh idirlín, na sonraí cearta a uaslódáil agus an pas bordála a chlóbhualadh. In ionad bheith ag útamáil leis an suíomh agus an trealamh, d'fhéadfása rud iomlán difriúil a dhéanamh. I mbeagán focal, ní mór don tomhaltóir glacadh anois le costais a bhí á n-iompar go dtí seo ag an táirgeoir. Léiríonn na costais chineachta seo nach bhfuil i ndioscúrsa an tsladmhargaidh ach sórt cur i gcéill. Íocann lucht labhartha mionteanga go laethúil as forlámhas na mórtheanga. Is costas ollmhór, seasta é an stró agus an deischostas a bhíonn i gceist agus na cainteoirí mionteanga ag iarraidh a ngnó a dhéanamh ina dteanga féin. Ní mór dúinn na costais sin a léiriú go beacht agus an bonn a bhaint ó reitric bhréagach an neamhspleáchais airgeadais. Is gníomh idéolaíoch teanga é costas a lua chun fáil réidh le hiolrachas teanga agus cultúir agus cleas eile chun cur leis na costais atá á n-iompar cheana féin ag cainteoirí na mionteanga.

An ceathrú heilimint atá ag teastáil uainn agus muid ag réiteach dioscúrsa nua do ghluaiseacht

na teanga ná athsmaoineamh a dhéanamh ar choincheap na háitiúlachta. De ghnáth, agus go mór mór i dtréimhse éiginnte, chorraitheach, shuaite is furasta cloí leis an rud áitiúil, an suíomh logánta mar thearmann, mar rud socair, mar sciath, mar bhalla cosanta in aghaidh thuilte uafara na nua-aimsearthachta. Is léir go bhféachtar go minic ar an nGaeilge sa tír seo mar thearmann den sórt sin. Samhlaítear an teanga fós mar fheiniméan atá ceangailte le miotas an oileáin iartharaigh. Gach uair beagnach a luaitear í ar na meáin chumarsáide Béarla, athraíonn an fuaimrian go holagón caointeach na bpíob uilleann agus taispeántar pictiúir fhileata de bhailte beaga báite le cúr na dtonnta agus solas ildaite na spéire fiáine. An dearcadh go dtí seo agus lucht labhartha na Gaeilge ag iarraidh na híomhánna sin a bhréagnú ná íomhánna eile a chur chun cinn a leagann an bhéim ar an nuaíocht agus ar an nua-aimsearthacht.

Ach smaoinigh nóiméad ar scríbhneoir Francach darbh ainm Xavier de Maistre a rugadh san ochtú haois déag. Scríobh sé leabhar a tháinig amach in 1794 dar teideal *Voyage autour de ma chambre*. Sa leabhar seo, a scríobhadh mar chuntas taistil, cuireann sé síos go mion ar a oilithreacht óna leaba go dtí an cófra agus ar aghaidh go dtí an folcadán agus an tolg. Baineann sé úsáid as teanga fhíordhrámatúil na

gcuntas taistil chun baol agus contúirt agus priacail an turais seo a chur in iúl go háiféiseach don léitheoir. Ar bhealach neamhghnách taispeánann de Maistre go bhfuil castacht agus doimhneacht ag baint leis na heachtraí is suaraí dár saol laethúil. Dhá chéad bliain ina dhiaidh, chuaigh François Maspero ar aistear eile. Fuair sé an traein a théann ó aerfort Charles de Gaulle go *Paris Gare du Nord*. De ghnáth maireann an turas thart ar dhaichead nóiméad. Ach chaith sé féin dhá mhí ar an turas sin, ag taisteal ón aerfort go lár na cathrach. Cad a tharla?

Deir sé sa leabhar a scríobh sé faoin aistear seo, *Les Passagers du Roissy Express* (1990), gur thuirling sé ag chuile stáisiún ar an mbealach agus gur chaith sé tréimhse sách fada i ngach uile áit. Fuair sé amach go raibh castacht agus iolrachas as cuimse ag baint leis na háiteacha a bhí go dtí sin mar fhásach uirbeach idir an t-aerfort gléineach, iar-nua-aoiseach agus an ghealchathair sheanda. Nuair a théann Maspero go mion isteach sa scéal tagann na céadta is na céadta scéal amach. Is ionann cúngú spáis agus fairsingiú intinne. An téarma atá cumtha agamsa don chur chuige seo seo ná *microspection* nó mioniniúchadh (Cronin 2012). Séard atá i gceist agam leis an gcoincheap seo ná modh anailíseach agus modh feidhmiúil a oibríonn ar leibhéal áitiúil agus a thaispéanann ní hamháin doimhne agus saibhreas na

háite ach an bealach nó na bealaí a cheanglaíonn an áit le háiteacha eile nó leis an domhan taobh amuigh.

Dhá shampla phraiticúla a léiríonn an cur chuige seo, logainmneacha agus ginealas. Tá tráidisiún láidir, ársa sa Ghaeilge, mar atá, an dinnseanchas, a leagann an bhéim ar thábhacht logainmneacha. Cad a léiríonn logainm, ainm áite? Ní hamháin scéalta agus miotais agus eolas faoin chomhthéacs tíreolaíoch (sin saibhreas inmheánach na háite) ach na hiarsmaí allúracha, coimthíocha atá le feiceáil nó le cloisteáil san ainm (sin saibhreas seachtrach na háite). Ar shuíomh idirlín faoi Shord Cholm Cille tugann an t-ealaíontóir Caoimghin Ó Croidheáin léargas dúinn ar an gceangal idir an baile agus na Lochlannaigh, muntir na Breataine Bige, na Gaeil agus na Sasannaigh trí anailís a dhéanamh ar na hainmneacha éagsúla a tugadh ar an áit. Maidir leis an nginealas, réimse staidéir a bhí tábhachtach riamh do na Gaeil agus cúram an ghinealaigh ar fhilí na scol, is léir go gcuireann an-chuid daoine spéis ann mar a thaispeáin an tsraith clár teilifíse, '*Who do you think you are?*' Rud a bhí follasach fiú ag leibhéal an duine aonair, chomh maith le castacht agus fiúntas stair clainne, go raibh an oiread sin teagmhálacha leis an gcoigríoch. Sa dá chás, logainmneacha agus ginealas, ní féidir fáil réidh leis an nGaeilge, teanga ár sinsir. Mar sin, is bealach é an mioniniúchadh seo chun

ceist na teanga a lonnú i gcroílár na héiceolaíochta polaitiúla.

An cúigiú heilimint atá le cur isteach sa phromhadán agus Turgnamh na hÉireann á ullmhú againn ná coincheap na poblachta. Nuair a d'fhoilsigh an *Irish Times* sraith alt faoi 'Renewing the Republic' (athnuachan na poblachta) in 2010, mar a dúramar cheana, ní raibh oiread is alt amháin ag cíoradh ról na Gaeilge i gcruthú nó in athmhúnlú na poblachta nua seo. In 2012 tháinig cnuasach aistí amach faoi stiúir Fintan O'Toole, leaseagarthóir an pháipéir chéanna. An teideal a bhí ar an gcnuasach ná *Up the Republic: Towards a New Ireland* ach arís níl oiread is abairt amháin ann faoi thábhacht na teanga mar chuid d'athghabháil an dóchais in Éirinn. Le linn chomóradh Eirí Amach na Cásca bhí tuilleadh á scríobh ar stair agus staid na poblachta ach níor smaoinigh mórán tráchtairí ar phríomhuirlis urlabhra agus samhlaíochta mhuintir an oileáin ar feadh na céadta bliain. Ní *New Ireland* a bhí ann ach athrá leibideach ar an seanphort aonteangach. Agus é ag caint mar gheall ar choincheap clasaiceach an tsaoránaigh i bhfealsúnacht na sean-Ghréagach, dúirt an fealsúnaí Éireannch Joseph Dunne an méid seo:

> The primary good[..]resided in the exercise of this very capacity to share in

> the project of self-rule. Citizenship then was demanding and to us its most obvious demand lay in the premium it required one to put on the public good even at the expense of one's private interests. (Dunne 2002: 75)

Conas féinriail a phlé nó leas éiceolaíoch an phobail a chur chun cinn sa chomhthéacs Éireannach gan foinse féinmhuiníne, tobar inspioráide agus an stór eolais atá le fáil sa Ghaeilge a chur san áireamh? Ní féidir poblacht nua ar bith a thógáil nó a chruthú i bhfásach frithchultúrtha an aineolais.Tosnaíonn 'Tuireamh na hÉireann' leis na línte brónacha:

> An uair smaoinim ar shaoithibh na hÉireann,
> scrios na dtíortha is díth na cléire,
> díothú na ndaoine is laighead a ngréithe
> bí mo chroíse im chlí dá réabadh.

Más sochaí inbhuanaitheach atá uainn ní foláir fáil réidh leis an Tuireamh agus tosnú athuair leis an Turgnamh.

Tráma

Ar an gcúigiú lá déag de mhí na Feabhra 2015 eagraíodh mórshiúl i mBaile Átha Cliath. Cuireadh i leith an rialtais easpa spéise agus patuaire maidir le forbairt agus cothú na Gaeilge. Ar an léirsiú seo, Dearg le Fearg, bhí an-chuid postaer agus plaiceard faoin mana clúiteach a luaitear le Tomás Dáibhís, Tír gan Teanga, Tír gan Anam. Níorbh é seo an dearcadh nó an tuairim a bhí ag an gComhairleoir Sinsearach d'Aire na Gaeltachta, Heather Humphreys. Ar an séú lá de mhí na Samhna 2014 scríobh sé meamram chuig oifigigh shinsearacha sa Roinn tar éis clár sealadach d'imeachtaí le haghaidh Chomóradh Éirí Amach na Cásca a chur timpeall. Maidir le seacht gceannteideal nó tosaíochtaí an Chláir (Gaolta, Daoine Óga, an Diaspora agus araile) mhol an comhairleoir go mbeadh an Ghaeilge ag bun an liosta díreach roimh Stampaí Poist. (McGee 2015). Nuair a foilsíodh an leagan Gaeilge de chlár na n-imeachtaí ar suíomh gréasáin oifigiúil de chuid an rialtais bhí sé do-léite agus do-thuigthe. Séard a bhí ann ná aistriúchan ciotach ó *Google Translate*. Ar ndóigh, nuair a scaipeadh an scéal mar gheall ar an ísliú céime don teanga, tháinig na gnáth-leithscéalta amach ó na foinsí oifigiúla. Ach cad is brí leis an naimhdeas nó leis an bhfaillí seo agus cén fáth go

bhfuil sé chomh coitianta sin i sochaí na hÉireann faoi láthair?

Agus í ag cur síos ar thráma idirghlúine na ndaoine a tháinig slán ón gcinedhíothú san Airméin luann an síocanailísí Janine Altounian an ceangal nimhneach idir ocras agus tost i meon an dreama seo:

> 'Il y aurait ainsi un lien étroit entre *mourir de faim* et *manquer de mots* pour se dire, pour se penser, pour nommer qui on est et signifier le monde dans sa propre langue' [Bheadh, mar sin, dlúthcheangal idir *bás a fháil den ocras* agus *na focail a bheith in easnamh* chun cur síos ar do chás féin, chun do mhachnamh féin a dhéanamh, chun tú féin a chur i láthair agus chun domhan lán de bhrí a chur ar fáil i do theanga (Altounian 2012: 149)

Is ionann Gorta agus ionsaí marfach ar dhá phríomhfheidhm na béalaireachta, an t-ithe agus an labhairt. Nuair a bhíonn feidhm amháin – an cothú – faoi bhagairt an scriosta ní haon ionadh go gcuireann sé seo isteach ar an bhfeidhm eile – an chumarsáid. Ag scríobh faoi stair a mhuintire i gCo. Mhaigh Eo, tugann Barry McCrea cuntas dúinn i *Languages of the Night* (2015), faoin gcliseadh teanga

sa chontae aimsir an Drochshaoil nuair a dhiúltaigh tuismí labhairt lena bpáistí ina dteanga féin agus iad scanraithe ag taibhse an ghanntanais bia, an t-ocras sin a dhruid béal na gcéadta mílte. Cad í an íomhá atá ag Gearóid Ó Tuathaigh agus é ag tabhairt léargas dúinn ar an athrú teanga san aimsir iar-Ghorta in Éirinn? 'The shame of the rising generation [...] may have been exacerbated by the horror of the Famine experience, where Irish was on the lips of the underclass dying in the ditches or heading to the poorhouse' (Ó Tuathaigh 2015: 35). Muna bhfuil na beola in ann bia a bhlaiseadh, is beag blas a bhíonn ag an teanga a bhlaiseann.

Smaoinigh arís ar an raic i dTuaisceart Éirinn roimh na toghcháin don Tionól. Cad a dúirt Arlene Foster mar gheall ar Acht na Gaeilge, "If you feed a crocodile it will keep coming back and looking for more' (Mallie 2017). De thaisme, gan dabht, agus ní d'aon ghnó, bhain sí úsáid as meafar an bhia, meafar an chothaithe ach mhúscail an ceangal seo idir ithe agus chumarsáid, bia agus briathra, fearg na náisiúnaithe sna Sé Chontae. Toradh amháin ar an dianocras anseo in Éirinn ná gur chuir sé dlús leis an athrú nó aistriú teanga a bhí ar siúl cheana féin, agus fillfimid níos déanaí ar ról an Ghorta Mhóir sa phróiseas seo. I bhfocail Uí Thuathaigh, 'the most fundamental social and cultural change in modern Irish history was the

shift from Irish to English as the vernacular of the population at large' (Ó Tuathaigh 2015: 7). Aistríodh an-chuid Éireannach ó theanga amháin go teanga eile ní hamháin in Éirinn ach thar lear freisin nuair a thréig siad teanga a muintire. Ní haon ionadh mar sin go raibh sampla de dhrochaistriúchán i gcroílár scéal an Chomórtha. Léirigh an t-aistriúchán ciotach an teannas a bhaineann le cúrsaí cuimhneacháin nuair a bhíonn éiginnteacht ann maidir le teanga agus cuimhní in Éirinn. Cé chomh fada agus is féidir leat dul i stair do phobail nó do mhuintire gan dul i ngleic le ceist seo an athraithe (nó an aistrithe) teanga? Scríobh an síocanailísí Ungárach Sandor Ferenczi nóta ina dhialann chliniciúil mar gheall ar dhaoine a thagann slán ó thubaiste. Is féidir le cuid den duine, dar leis, bás a fháil ach má thagann cuid eile den duine slán ón tráma, fágadh é agus 'bearna ina chuimhne' nó 'bearna ina phearsantacht' (luaite in Altounian 2012: 39). Is ionann an teanga nach labhraítear a thuilleadh agus cathair thoirmiscthe atá faoi léigear ag mothúcháin dhiúltacha an ghortaithe. I gcás tharchur idirghlúine an tráma, gné amháin den leanúnachas trámaithe ná fanacht dílis do thost an dísheallbhaithe, an *great silence* a luaigh Seán de Fréine na blianta fada ó shin (de Fréine 1978).

Ar ndóigh, ní féidir locht an athraithe a chur an nGorta Mór amháin mar a chruthaigh Gearóid Ó

Tuathaigh sa leabhar *I mBéal an Bháis: The Great Famine and Language Shift in Nineteenth-Century Ireland* (2015). An córas oideachais, cúrsaí imirce, ról na hEaglaise, an córas dlí, fórsaí geilleagracha, bhí siad go léir beagnach ag feidhmiú i gcoinne na Gaeilge roimh thubaiste an Ghorta. Ar ndóigh, mar a mhaíonn Nicholas M. Wolf in *An Irish-Speaking Island: State, Religion, Community and the Linguistic Landscape, 1770-1870* (2014), bhí an t-aonteangachas Gaeilge agus an dátheangachas i bhfad níos forleithne ná mar a ceapadh uaireanta anuas go dtí aimsir an Ghorta, agus sa leabhar úd Ó Chéitinn go Raifthearaí (2011) rinne Vincent Morley cur síos bríomhar ar théagar agus ar bheocht an chultúir Ghaelaigh ag tús an naoú haois déag. Dála Bhreandáin Uí Bhuachalla ag caint faoi na manaigh i Lováin in *Aisling Ghéar* (1996), níl Wolf ná Morley sásta glacadh le meafar bitheolaíoch an bháis, stair na teanga agus a lucht labhartha de shíor á plé faoi scáth an tuama. Ní raibh drogall ar bith ar an-chuid cainteoirí san ochtú agus sa naoú haois déag an teanga a labhairt agus a láimhseáil in an-chuid réimsí den saol poiblí. Léiríonn an fhianaise seo go léir cé chomh láidir agus chomh torthúil is a bhí an teanga roimh aimsir an Drochshaoil, nach raibh aon chinnteacht ann gurbh í an bás a bhí i ndán di ach ag an am céanna, is léiriú í freisin ar cé chomh

tubaisteach is a bhí tráma an Ghorta.

Go deimhin, roimh 1845 bhí an teanga á tréigean ag cainteoirí áirithe ach níl an claonadh seo inchurtha le meath ollmhór aimsir an Ghorta. Mar a léiríonn Ó Tuathaigh:

> The cultural exchange or intercourse between the two vernaculars – given the realities of power – cannot be judged as anything other than unequal and grossly asymmetrical. From the time of the Conquest, the law, politics, administration, indeed the entire State apparatus was conducted through English: legal documents, public announcements, proclamations and the administration of justice were all in English (Ó Tuathaigh 2015: 24)

Déanann Patricia Palmer mionchur síos ar an bhforéigean fíochmhar a cuireadh i bhfeidhm chun an teanga agus lucht a labhartha a dhíbirt as an saol poiblí sna leabhair *Language and Conquest in Early Modern Ireland* (2001) agus *The Severed Head and the Grafted Tongue: Literature, Translation and Violence in Early Modern Ireland* (2011). Sampla amháin ná an sliocht seo as *A General Rehersal of Warres* (1579)

le Thomas Churchyard. Fear bolscaireachta ba ea Churchyard agus tugann sé cuntas mórtasach dúinn de straitéis amháin a bhí in úsáid ag ceannasaí míleata Túdarach. An nós a bhí ag an bhfear airm seo ná:

> that the heddes of all those (of what sort soeuer thei were) which were killed in the daie, should bee cutte of from their bodies, and bought to the place where he incamped at night: and should there bee laied on the ground, by eche side of the waie leadying into his owne Tente: so that none could come into his Tente for any cause, but commonly he muste passe through a lane of heddes, which he used *ad terrorem,* the dedde feelying nothying the more paines thereby: and yet did it bryng greate terrour to the people, when thei sawe the heddes of their dedde fathers, brothers, children, kinsfolke, and freendes, lye on the ground before their faces, as thei came to speake with the saied Colonell (Q.iii.v).

An fear a bhí freagrach as an sléacht ná an Ridire Humprey Gilbert, leasdeartháir leis an Ridire Walter Raleigh. Scoláire ilteangach, sáraistritheoir ón Spáinnis, iarscoláire de chuid Eton agus céimí de

chuid Oxford, fear a bhí bródúil as an oideachas a fuair sé sna daonnachtaí agus a luaitear san eagrán is déanaí den *Oxford Dictionary of National Biography* mar "man of considerable intellectual sophistication" (Rapple 2004). Cinnte, bíonn an baol ann i gcónaí agus muid ag tarraingt aird ar stair fhuilteach an Choncais go bhfuilimid ag glacadh go rófhurasta leis an dearcadh a cháin Liam Kennedy sa leabhar úd, *Unhappy the Land* (2015) leis an fo-theideal amhrasach, *The Most Oppressed People Ever, the Irish?*

Ach ón gConcas ar aghaidh is deacair a shéanadh go raibh cúrsaí teanga fite fuaite le hoidhreacht an fhoréigin agus fíricí na héagothroime. Go hachomair, ní féidir a rá gur bhain an tráma teanga le haon eachtra eisceachtúil amháin ach le sraith thubaisteach d'imeachtaí éagsúla. Ach cén fáth, mar sin, agus roinn síceolaíochta i mbeagnach gach aon ollscoil in Éirinn, nach ndearnadh oiread is staidéar cuimsitheach amháin ar 'the most fundamental social and cultural change in modern Irish history'? Cén fáth go raibh cúrsaí teanga chomh himeallach sin i gcomóradh an Ghorta Móir i 1995? Fanaimid inár dtost faoin tost. Nó an bhfuil sé sin fíor? An bhfuil comharthaí ann sa tsochaí a léiríonn foréigean folaithe an tráma, ó bhrúidiúlacht ranganna Gaeilge i scoileanna áirithe tar éis bhunú an tSaorstáit a bhí díreach mar an gcéanna leis an íde choirp sa

seanchóras go dtí na hargóintí teasaí faoi cheist na teanga a bhriseann amach am ar bith a luaitear í sa saol poiblí nó go deimhin príobháideach (cóisir, teach tábhairne)? Ciontacht, fearg, míshuaimhneas, féinfhuath, ní minic a dhéantar dearmad ar na mothúcháin seo agus cúrsaí Gaeilge i gcroílár an aonaigh. Ar smaoinigh ailtirí na hAthbheochana a ndóthain ar an dochar síceolaíoch a bhaineann leis an gceangal idir foréigean fisiciúil agus eacnamaíoch agus athrú nó aistriú teanga?

An sprioc atá agamsa, ní hé cur síos stairiúil a sholáthar ar athrú teanga in Éirinn ach cúpla ceist a ardú mar gheall ar ghnéithe síceolaíocha an athraithe agus comhthéacs na héiceolaíochta á bplé againn. Más fíor go bpléitear sách minic forbairt stairiúil na teanga nó cúinsí soctheangeolaíocha a bhaineann le cleachtadh comhaimseartha na teanga is fíorannamh a luaitear oidhreacht shíoc-chultúrtha an athraithe teanga. Ba mhaith liom mar sin cúpla gné d'oidhreacht an tráma a phlé.

Ag scríobh faoi thaighde Ruth Ley ar choincheap an tráma deir Warwick Anderson, Deborah Jenson agus Richard C. Keller an rud seo a leanas:

> In Ruth Ley's problematization of the character of trauma as mimetic trauma or "traumatic mimesis", the victim almost

> hypnotically identifies with the aggressor and is open to imitation and therefore repetition, in different subject positionings of traumatic experience' (Anderson, Jenson agus Keller 2011: 14)

Muna bhfuil meas ag an lucht ceannais ar an teanga, ní fada go mbeidh rian an díspeagtha sin le feiceáil i meon na ndaorchlainne. Sampla amháin i measc na mílte ná an méid a scríobh Micheál Ó Raghallaigh sa naoú haois déag ag gabháil leithscéil mar gheall ar chaighdeán a chuid scríbhneoireachta:

> A léitheoir ionúin, ná tabhair aithis ná milleán orm trí olcas an scríbhinn atá sa leabhar seo. Óir is fear ceirde mé agus is in aimsir dhíomhaoin a rinne mé an cnuasach seo. Dá bhrí sin, an tan a chonacthas dom nach raibh duine de lucht mo chomhaimsire ag cnuasach a bheag de shaothar ár bhfilí, mheas mé gan iad a ligean ar fán. Ach is beag an tairbhe iad a chosaint, óir is náireach lenár n-aos óg teanga a fhogluim. Dá bhrí sin ní bhia focal Gaeilge sa ríocht i gceann céad bliain eile má lenaid an nós atá acu lem'chuimhne féin (Ó Fiannachta 1995: 98).

Cé hiad na heochairfhocail sa réamhrá seo? 'Aithis', 'milleán', 'olcas', 'díomhaoin', 'náireach', léiriú ar an bhféinfhuath, ar an ngortú síceach a leanann tráma i sochaí ar bith. Cinnte, ní raibh an ceart ag Ó Raghallaigh, tá Gaeilge fós á labhairt ní sa ríocht ach sa Phoblacht ach cad mar gheall ar an miméis thrámaithe, oidhreacht na mothúcán diúltach? Níos gaire dúinn, foilsíodh alt san *Irish Times* leis an scríbhneoir agus iriseoir Rosita Boland. An teideal a bhí ar an alt ná 'Can anybody truthfully say that Irish is a necessary language?' Cháin sí go géar an easpa cumais a bhí aici sa teanga tar éis na blianta a chaitheamh ag foghlaim Gaeilge agus an méid airgid a chaith an rialtas ar aistriú cáipéisí. An ghné ba spéisiúla, áfach, den alt ná an friotal a bhí á úsáid aici, 'The disgrace, as I see it, is being forced by the State to study a compulsory language for which I have no aptitude, absolutely no interest in, and no choice about throughout my entire school career' (Boland 2016). 'Disgrace', 'compulsory', 'no aptitude', 'no interest', 'no choice', seo iad na focail agus na frásaí a chuireann in iúl don léitheoir an naimhdeas atá ag an scríbhneoir don Ghaeilge. Cinnte, is beag an 'choice' nó an rogha atá ag an duine a labhraíonn Gaeilge an teanga a úsaid in an-chuid réimsí den saol poiblí in Éirinn, ach níl fonn orm leanúint ar aghaidh leis an díospoireacht sin.

B'fhearr liom cloí le focail an ailt féin mar chomhartha ar an miméis trámaithe. D'fhéadfadh, mar shampla, Micheál Ó Raghallaigh na focail cheannann chéanna a úsáid chun cur síos ar a chás féin, 'compulsory', 'disgrace', 'no choice'. Is léirmhíniú alt Boland ar an miméis, an 'imitation', a thagann as tráma, fiú amháin i láimhseáil na teanga féin mar a mhaíonn Ruth Ley. Sampla an-mhaith den athrá trámaithe ná an focal 'no', trí uair i ndiaidh a chéile in aon abairt amháin. Is suntasach, mar sin, an méid a scríobhann Boland maidir le suíomh na Gaeilge sa chóras scolaíochta, 'It's confusing to load a school subject with so many things that have nothing to do with education, and everything to do with the wider and highly emotive national question of identity' (Boland 2016). Taispeánann an t-údar, go neamh-chomhfhiosach gan amhras, an buntráma a bhaineann le ceisteanna teanga nuair a dhéanann sí tagairt do 'highly emotive national question of national identity'. Cén fáth go mbeadh mothúcháin láidre i gceist maidir le múineadh na Gaeilge agus doicheall roimh í a fhoglaim agus gan an deacracht chéanna a bheith ann maidir le múineadh agus foghlaim an Bhéarla? Go mór mór, mar a dhearbhaíonn an scríbhneoir, go bhfuil an Béarla mar dhlúthchuid dá hionnanas náisiúnta mar Éireannach. Toisc nach n-aithnítear oidhreacht

mhothúchánach tráma an athraithe teanga níl aon fhuascailt ón athrá, ón aithris, ó na frámaí tagartha diúltacha atá mar chrann taca do dhíospóireachtaí poiblí faoi cheist na teanga in Éirinn.

Maidir le stair na n-impireachtaí sa naoú haois déag agus san fhichiú haois cloistear go minic scéalta mar gheall ar eachtraí a tharla in impireacht na Breataine nó impireacht na Fraince, agus is beag an cur amach atá ag an ngnáth-Éireannach ar stair nua-aimseartha na himpireachta a bhí ag na hOllanaigh san Indinéis agus i dtíortha nach í. Fear amháin agus a lán scríofa aige faoi stair na síciatriachta sna coilíneachtaí Ollanacha ná Hans Pols. Ag déanamh cur síos ar an meon a bhí ag na hOllanaigh i leith na mbundúchasach sa chuid sin den domhan, deir sé:

> In the colonial Dutch East Indies, there was, on the one hand, an outspoken disdain for the lazy, emotional, childish and uneducated native who could not possibly benefit from Western education; on the other hand, there was also an exaltation of Javanese Eastern spirituality and high culture, which had existed for thousands of year and had been passed down from generation to generation. (Pols 2011: 145)

Ba mhian liom díriú isteach ar an gcéad chuid den mheon sin, is é sin, íomhá na mbundúchasach mar dhream leisciúil, mothúcánach, páistiúil, aineolach. Ní nach ionadh, bhí an meon sin coitianta ní hamháin i measc na nOllanach ach i measc na nIartharach i ré na n-impireachtaí. Scríobh Jakob Venedey, polaiteoir agus iriseoir, leabhar taistil *Irland* (1844) faoi chuairt ar Éirinn dhá bhliain roimh an nGorta Mór in 1843. Bhí an-mheas aige ar an bhfeachtas a bhí ar siúl ag Dónall Ó Conaill ar son an reipéil agus is beag bá a bhí aige le rialtas na Corónach. Ach fós féin, ní raibh sé in ann éalú ó na réamhthuairimí smolchaite maidir le carachtar na nÉireannach. Ag ollchruinniú de chuid an Chonallaigh in Áth Luain molann sé bua na cainte atá ag Ó Conaill, caint a thugann ugach agus inspioráid dá mhuintir. I bhfianaise an Ghearmánaigh:

> Is fíor-Éireannach pearsantacht [an Chonallaigh] nó ba chóir dom a rá pearsantacht a chuid óráidí. Do bheadh sé fánach loighic leanúnach a aimsiú iontu, ach b'fhéidir go bhfuil sé sin fíor i gcás an-chuid cainteoirí cáiliúla [...] Nuair a labhraíonn sé, bíonn seisean agus an lucht éisteachta mar aon duine amháin; ní féidir leis an Éireannach maoithneach smacht a

choimeád ar a chuid mothúchán (luaite i Bourke 2011: 434)

Ar ndóigh, ní haon eisceacht é Venedey agus cuireadh síos go minic ar na hÉireannaigh sa thréimhse sin mar pháistí saonta, iad tugtha do bhrionglóidí agus do liriciúlacht ach ar an drochuair, cosúil le páistí dána ag easaontú go bagrach leis an údarás cuí. I gclann uilíoch na daonnachta, bhí cuid den chlann níos forbartha, níos aibí ná an chuid eile agus bhí dualgas agus cúram na ceannasaíochta ag luí go nádúrtha ar an dream lánfhásta. Le coincheap na haibíochta sa tsíciatriacht choilíneach faightear an ceangal idir teoiric bhitheolaíoch na héablóide agus teoiric shíceolaíoch an chiníochais. Ach an bhfuil oidhreacht na smaointeoireachta seo le feiceáil fós agus cén bhaint atá aici le ceist an tráma agus an athraithe teanga? Ag scríobh le déanaí san *Irish Times* faoi chuairt eile, cuairt a thug an Ginearál Aoun, Uachtarán na Liobáine, ar an bhFhrainc, rinne Lara Marlowe comparáid idir proiséas na síochána sa Liobáin agus i dTuaisceart Éireann, 'Like the protagonists of the Troubles in Northern Ireland, the Lebanese have matured' (Marlowe 2017). D'fhás muintir na Sé Chontae agus muintir na Liobáine aníos agus anois tig linn glacadh leo mar dhaoine lánfhásta. Bíonn tuismí réasúnta an oird

dhomhanda ag fanacht go foighneach le deireadh a chur le suaitheadh na páistiúlachta. Nuair a chuireann Hugh Linehan an cheist, 'Is there anything more tedious than the Irish-language debate' (Linehan 2017), cloismid cinnte guth an duine stuama, aibí, réasúnta atá bréan de bheith ag éisteacht leis na páistí ag spochadh as a chéile sa seomra suite. Níl an dara rogha aige ach údarás na haibíochta a chur in iúl don mhuirín drochbhéasach fiú muna bhfuil sé dóchasach faoi thorthaí a chuid iarrachtaí, 'there's more reason than ever to take a break from the rows, the antagonisms, and the clashes and for once have a sensible debate on the matter. Don't hold your breath though'. Réamhthuiscint amháin atá taobh thiar den dearcadh seo ná gur féidir na mothúcháin maidir le cúrsaí teanga a scaradh amach ó na fíricí. Ón 'highly emotive question of national identity' go 'sensible debate on the matter' is léir go bhfuil na mothúcháin pháistiúla de shíor ag bagairt ar neodracht aibí an tráchtaire aonteangaigh.

Ach baineann imeallú na mothúchán i leith na teanga ní hamháin le cur faoi chois thráma an athraithe teanga ach le hidéolaíocht atá go láidir i gcultúr lucht labhartha an Bhéarla. Chum an socheolaí Boaventura de Sousa Santos an téarma *epistemicide* chun cur síos ar an nós atá ag eolaíocht an Iarthair bealaí eile chun teacht ar eolas a dhíspeagadh nó a scriosadh. Maidir

le cúrsaí dlí, cúrsaí leighis, cúrsaí timpeallachta níl ach an t-aon bhealach amháin chun an fhírinne a aimsiú is é sin cur chuige eolaíochtúil an Iarthair. Maíonn Karen Bennett go bhfuil baint ag an bproiséas seo le ceist na teanga freisin. Féach ar an mbrú leanúnach ar lucht léinn ar fud an domhain a chuid saothair a scríobh agus a fhoilsiú i mBéarla. Níl seasamh ar bith ag saothar i rangú idirnáisiúnta muna bhfuil sé foilsithe in iris idirnáisiúnta agus ciallaíonn 'idirnáisiúnta' i gcónaí, iris trí mheán an Bhéarla. Ach cén sort Béarla atá ar fáil sna hirisí acadúla? Go deimhin, séard atá ar fáil ná teanga atá fréamhaithe go smior i réabhlóid eolaíochta an tseachtú haois déag. I bhfianaise Bennett tá an EAD (English as Academic Discourse) seo fite fuaite leis an imeallú a rinneadh sa tréimhse sin ar eolas na scolaíochta agus léann scoláirí an Renissance a bhí bunaithe ar théacsanna agus ar lárnacht an fhocail scríofa. D'fhás deighilt idir an t-eimpíreachas Angla-Shacsanach ar thaobh amháin agus fealsúnacht na Mór-Roinne ar an taobh eile, an heirméineotaic, an t-idéalachas Gearmánach, an fheiniméaneolaíocht, an t-eiseachas, an structúrachas agus an t-iarstructúrachas:

> In the first case, "reality" is believed to precede language, which means that it is perceived in much the same way by

> everyone, irrespective of linguistic or cultural background; hence, language serves merely to name things that have an objective existence in some extralinguistic domain (a philosophy known as *linguistic realism*). The second, on the other hand, holds that humans learn about external reality through the categories set up by their mother tongue; thus our entire experience of the extralinguistic world is not only mediated but actively constructed by culture (a philosophy known today as *constructivism*). (Bennett 2015: 12)

Ceann de na deacrachtaí atá ag lucht labhartha na Gaeilge ná go bhfuil siad ag déileáil le hidéolaíocht teanga atá i mbarr a réime le proiséas an domhandaithe, idéolaíocht a cheileann leithleachas teanga, a mhaíonn nach bhfuil ann i gcás urlabhra ar bith ach uirlis thrédhearchach chun fíricí an nádúir a chur faoinár mbráid.

Agus muid ag caint faoi stair, cothú, dul chun cinn na Gaeilge, ní féidir neamshuim a dhéanamh de na cúinsí sochtheangeolaíocha a chuidigh agus atá ag cuidiú fós le meath na teanga. Ach is dóigh liom go mba chóir níos mó béime a chur ar an oidhreacht shíceolaíoch a bhaineann le cúrsaí teanga in Éirinn,

mar cloistear rian na hoidhreachta sin in an-chuid conspóidí mar gheall ar an teanga. Cuireadh agallamh le déanaí ar Éireannaigh a bhí ina gcónaí sa Chatalóin ag iarraidh a fháil amach cén tuairim a bhí acu maidir le neamhspleáchas na Catalóine. Fear amháin ó Bhaile Átha Cliath, John Kearns, ní raibh sé rófhabhrach do chúis an neamhspleáchais ach d'admhaigh sé go raibh an-mheas aige ar mhuintir na Catalóine agus an dúil a bhí acu ina dteanga féin, 'I wish Irish people were as proud of their language as the Catalans' (Hedgecoe 2017). Nach bhfuil sé in am dúinn an cheist sin a fhreagairt agus dul níos doimhne isteach sa fíor- 'Hidden Ireland', is é sin a rá, stair fholaithe an díshealbhaithe shícigh.

Ar ndóigh, ceist a cheanglaíonn stair na tíre agus todhchaí na sochaí ná ceist na héiceolaíochta. Thug George Petrie agus tráchtairí nach é faoi deara go raibh titim shuntasach i líon na sean-nósanna a bhí á gcleachtadh in Éirinn (Mac Suibhne). Gan amhras, bhí an-chuid de na nósanna sin nasctha le hómós cuí a thabhairt don timpeallacht agus do stair na timpeallachta. Bhí an tréigean seo mar chuid de phroiséas an nua-aimsearthachta a bhí forleathan sa tréimhse sin ar fud na hEorpa. Anois agus muid ag druidim le deireadh na ré sin a bhí dall ar an tubaiste leanúnach a bhí ar siúl in an-chuid réimsí den timpeallacht, ba chóir athbhreithniú a dhéanamh

ar oidhreacht an athraithe teanga maidir leis an dearcadh nó leis na luachanna atá ag an bpobal a mhaireann ar oileán na hÉireann. Déantar tagairt go mion minic faoi láthair don tráma éiceolaíoch a thagann ón scoilt mharfach idir an cine daonna agus an domhan nádúrtha timpeall orthu. (Morton 2018: 67). Deir scríbhneoirí áirithe go mbaineann an scoilt seo leis an meon i dtraidisiún áirithe sa Chríostaíocht a mhaíonn go bhfuil an domhan neamhdhaonna faoi cheannas an duine. Deir scríbhneoirí eile go bhfuil an locht ar Descartes agus agus ar aiséirí Réabhlóid na hEolaíochta a thug príomháit don intinn agus don intleacht mar uirlis chun smacht a fháil ar an nádúr. Rud amháin atá faoi an-chuid díospóireachtaí poiblí in Éirinn ná go mbaineann siad le réadmhaoin. Ciallaíonn géarchéim sa tír de ghnáth, géarchéim tithíochta, géarchéim réadmhaoine. Bíonn plé ann faoi shealbhú agus faoi shealúchas ach dada faoi áitreabh, faoi inmharthacht, faoi bheith i do chónaí go feasach in áit faoi leith? An bhfuil ceangal mar sin idir an tráma éiceolaíoch agus an tráma teanga? An bhfuil deacrachtaí ag pobal tionchar agus cruth na timpeallachta a thuiscint toisc go bhfuil sí faoi cheilt i dteanga eile?

Cinnte, níl sé furasta uaireanta fianaise na síceolaíochta nó na sícanailíse a láimhseáil. Ach ní dóigh liom gur féidir linn a thuilleadh agus muid ag

dul i ngleic leis na dúshláin mhóra éiceolaíocha atá romhainn neamhshuim a dhéanamh den 'mhórthír dhorcha' a luaigh an teoiriceoir Ostarach Sigmund Freud. Níl aon réiteach ar scéal na timpeallachta agus ní bheidh muna bhfuilimid sásta tábhacht na Gaeilge a aithint mar bhunchloch d'aon pholasaí réadúil as seo amach agus cothrom na féinne a thabhairt don chomhthéacs stairiúil agus síceolaíoch ina iomlán.

Tagairtí

Altounian, Janine (2012) *De la cure à l'écriture: L'élaboration d'un héritage traumatique*, Páras: PUF.

Anderson, Warwick, Jenson, Deborah agus Keller, Richard C. (eag.) *Unconscious Dominions: Psychoanalysis, Colonial Trauma, and Global Sovereignty*, Durhan (NC): Duke University Press.

Beck, John agus Davenport, Thomas (2001) *The Attention Economy: Understanding the New Currency of Business*, Cambridge (MA): Harvard Business School.

Bennett, K. (2015), 'Towards an epistemological monoculture: Mechanisms of epistemicide in European research publication', 9-35, in Plo Alustrué, Ramón and Pérez-Llantada, Carmen (eag), *English as a Scientific and Research Language*, Beirlín: De Gruyter Mouton.

Boland, Rosita (2016) 'Can anybody truthfully say that Irish is a necessary language', *The Irish Times*, 30 Bealtaine.

Bourke, Angela agus Negra, Diane (2019) *An Ghaeltacht, An Ghaeilge, An Béaloideas agus Cruthaitheacht na nDaoine*, Baile Átha Cliath: Acadamh Ríoga na hÉireann.

Bourke, Eoin (2011) *"Poor Green Erin": German Travel Writers' Narratives on Ireland from Before the 1798 Rising to After the Great Famine*, Berlin: Peter Lang.

Cronin, Michael (2012) *The Expanding World : towards a politics of microspection*, Washington: Zero Books.

Darmody, Merike agus Daly, Tania (2015) *Dearcadh an Phobail i leith na Gaeilge ar oileán na hÉireann*, Baile Átha Cliath: ESRI/Foras na Gaeilge.

de Fréine, Seán (1978) *The Great Silence*, 2nd ed., Cork: Mercier.

De Maistre, Xavier (1959) *Voyage autour de ma chambre*, Páras: Laffont.

de Paor, Louis (2018) *Ag Caint leis an Simnê: Dúshlán an Traidisiúin agus Nualitríocht na Gaeilge*, An Spidéal: Cló Iar-Chonnacht.

Dunne, Joseph (2002) 'Citizenship and Education : A Crisis of the Republic?', Peadar Kirby, Luke Gibbons agus Michael Cronin (eag.), *Reinventing Ireland: Culture, Society and the Global Economy*, Londain: Pluto Press.

Fogarty, Pádraig (2017) *Whittled Away: Ireland's Vanishing Nature*, Corcaigh: Collins.

Grin, François (2004) 'On the Costs of Cultural Diversity', in van Parijs, P. (ed.) *Cultural Diversity versus Economic Solidarity*, Bruiséal: De Broeck Université, 189-202.

Hamilton, Sorcha (2019) 'How to make your eco-lunches more friendly', *The Irish Times Magazine*, 21 Meán Fómhair, 11.

Harrison, Bernice (2012) 'A portrait of a comedian in a deadly serious situation' *The Irish Times*, 10 Márta.

Hedgecoe, Guy (2017) 'Irish in Catalonia: Scepticism and idealism as independence vote looms', *The Irish Times*, 27 Meán Fómhair.

Joyce, P.W. (1883) *The Origin and History of Irish Names of Places*, iml. 2, Baile Átha Cliath: Gill.

Judet de la Combe, Pierre agus Wismann, Heinz (2004) *L'Avenir des languages: repenser les humanités*, Páras: Éditions du Cerf.

Judt, Tony (2010) *Ill fares the land*, Londain: Penguin.
Kennedy, Liam (2015) *Unhappy the Land: the Most Oppressed People Ever, the Irish?*, Baile Átha Cliath: Merrion Press.
Leerssen, Joep (1996) *Mere Irish and Fíor-Ghael*, Corcaigh: Cló Ollscoile Chorcaí.
Linehan, Hugh (2017) 'Is there anything more tedious than the Irish language debate?', *The Irish Times*, 2 Meán Fómhair.
Mac Brádaigh, Frainc (2010) *Cá bhfuil Walsingham?* Baile Átha Cliath: Foilseacháin Ábhair Spioradálta.
Mac Suibhne, Breandán (2018) *The End of Outrage: Post-Famine Adjustment in Rural Ireland*, Oxford: Oxford University Press.
Mallie, Eamonn (2017) 'How Arlene Foster helped nationalism find its teeth', *The Irish Times*, 4 Meán Fómhair.
Marlowe, Lara (2017) 'Lebanese president's French visit marks new phase in relations', *The Irish Times*, 25 Meán Fómhair.
McCrea, Barry (2015) *Languages of the Night: Minor Languages and the Literary Imagination in Twentieth-century Ireland and Europe*, New Haven: Yale University Press.
McGee, H. (2015) 'Gaeltacht department sought relegation of Irish in 1916 event', *The Irish Times*, 9 Feabhra.
Maspero, François (1990) *Les Passagers du Roissy Express*, Páras: Seuil.
Morton, Timothy (2018) *Being Ecological*, Londain: Pelican.

Morley, Vincent (2011) *Ó Chéitinn go Raiftearaí: Mar a cumadh stair na hÉireann*, Baile Átha Cliath: Coiscéim.

Ní Dhroma, Máire (2019) 'Na Prátaí Dubha', *Poethead*. Ar fáil ar líne: https://poethead.wordpress.com/2010/08/06/amhran-na-bpratai-dubha-maire-ni-dhroma-trans-michael-coady-pir.

Ó Cuív, Brian (1989-90) '*Dinnseanchas* - The Literary Exploitation of Irish Place Names', *Ainm: Bulletin of the Ulster Place-Name Society*, 4, 90-106.

Ó Fiannachta, Pádraig (1995) (eag.) Ómós do Eoghan Ó Comhraí, An Daingean: An Sagart.

Ó Laoire, Lillis (2003) '"...d'imeodh an aois díom": Tír na nÓg agus Talamh na hÉireann', 39-68 in McLeod, Wilson agus Ní Annracháin (eag.) *Cruth na Tíre*, Baile Átha Cliath: Coiscéim.

Ó Maolalaigh, Roibeard (2003) 'Tírdhreach na nGael: Logainmneacha agus an tSamhlaíocht', 195-243, in McLeod, Wilson agus Ní Annracháin (eag.) *Cruth na Tíre*, Baile Átha Cliath: Coiscéim.

O'Sullivan, Kevin (2019) 'Ireland has third highest emissions of greenhouse gas in EU', *The Irish Times*, 26 Lúnasa. Ar fáil ar líne: https://www.irishtimes.com/news/environment/ireland-has-third-highest-emissions-of-greenhouse-gas-in-eu-1.3998041.

O'Toole, Fintan (2009) *Ship of Fools: How Stupidity and Corruption Sank the Celtic Tiger*, Londain: Faber and Faber.

O'Toole, Fintan (2012) *Up the Republic: toward a new Ireland*, Londain: Faber & Faber.

Ó Tuama, Seán (1992) 'Ómós Áite: a rian ar scríbhneoirí in Éirinn", *Comhar*, 51, 5, 176-185.

Ó Tuathaigh, Gearóid (2015) *I mBéal an Bháis: The Great Famine and Language Shift in Nineteenth-Century Ireland*, Hamden (CT): Quinnipiac University Press.

Pols, Hans (2011) 'The Totem Vanishes, the Hordes Revolt: A Psychoanalytic Interpretation of the Indonesian Struggle for Independence', 141-166, in Anderson, Warwick, Jenson, Deborah agus Keller, Richard C. (eag.) *Unconscious Dominions: Psychoanalysis, Colonial Trauma, and Global Sovereignty*, Durhan (NC): Duke University Press.

Proinsias (2016) *Laudato Si': Aitheasc Aspalta ar ár bhfreagracht i leith na timpeallachta*, aistr. Pádraig Ó Fiannachta, Má Nuad: An Sagart.

Rapple, Rory (2004) 'Gilbert, Sir Humphrey (1537-1583), explorer and soldier', *Oxford Dictionary of National Biography*, Oxford: Oxford University Press. Ar fáil ar líne: https://www.oxforddnb.com/search?q=Gilbert%2C+Sir+Humphrey+%281537–1583%29%2C+explorer+and+soldier.

Reynolds, Burnam W. (2012) *Columbanus: Light on the Early Middle Ages*, Londain: Pearson.

Thurneysen, Rudolph (1917) 'Morands Furstenspiegel', *Zeitshcrift für Celtische Philologie*, 9, 56-106.